STATISTICS EXPLAINED

A Computer Assisted Guide to the Logic of Statistical Reasoning

Howard S. Hoffman

Professor of Psychology
Bryn Mawr College
Bryn Mawr, Pennsylvania 19010

UNIVERSITY PRESS OF AMERICA

LANHAM • NEW YORK • LONDON

University Press of America,® Inc.

4720 Boston Way
Lanham, MD 20706

3 Henrietta Street
London WC2E 8LU England

Printed in the United States of America

Library of Congress Cataloging in Publication Data
Hoffman, Howard S., 1925-
Statistics explained.

1. Statistics—Computer assisted instruction.
I. Title.
QA276.4.H64 1985 519.5'07 85-15724
ISBN 0-8191-4894-6 (alk. paper)
ISBN 0-8191-4895-4 (pbk. : alk. paper)

All University Press of America books are produced on acid-free paper which exceeds the minimum standards set by the National Historical Publications and Records Commission.

Preface

This book represents an effort to explain the logic of statistical reasoning by showing how its otherwise puzzling formulae and procedures arise from a limited number of relatively straightforward basic propositions. The computer exercises that accompany this book are designed to enable you to explore these propositions and procedures in a direct empirical fashion. In many respects the task of teaching statistics is something of a challenge because most of the concepts of statistics are best expressed in purely mathematical terms and the majority of students do not have the mathematical background or sophistication to deal with such an approach. For this reason it is common practice in many statistics texts and courses to ask the student to accept certain formulae and propositions as given and to suggest that the interested student must go to other sources to discover their basis. That is not the approach of this book. Instead, I have tried to explain the several formulae and propositions in such a way as to lay bare their logical base by guiding the student through a series of small but intuitively reasonable conceptual steps. I have recognized, however, that there are several inherent dangers in presenting statistics in this kind of simplified fashion. Perhaps the most important is that one runs the risk of swamping the student with a mass of irrelevant detail. In an effort to avoid that problem, I have tried, wherever possible, to keep to basics and to avoid lengthy discussions of issues that often amount to little more than matters of taste. For example, many statistics texts devote a good deal of attention to the elaboration of various rules for graphically portraying data. While such rules may be helpful to some students in some instances, most students can recognize that in the long run how one portrays a given set of data is largely determined by the message one wishes to convey.

Another danger, related to the first, arises when one relies too heavily on real data to elaborate a given procedure. For instance, it is not uncommon to use examples from published pharmacological research to explain how one goes about an analysis of variance. Unfortunately, while such illustrations are sometimes helpful they often force students to digest large masses of unwieldy numbers and the require one to learn a great deal more about drugs, doses and days than one may care to know. In an effort to avoid this problem, I have tried, whenever possible, to keep the illustrative examples simple and to use numbers that are easy to assimilate. Hopefully the gains in comprehension will compensate for the sacrifices in realism.

Finally, there is the danger that things can be made too simple. Students sometimes find that when material is presented sequentially and simply they can easily proceed from one step to the next without ever really putting it together and deriving a clear picture of the subject as a coherent ensemble. That is, a student will often understand each of the parts of a given argument or illustration without actually attaining an integrated view of the principle being elaborated. There is, fortunately, a prophylaxis that can be applied in these circumstances. It derives from the essential truth that one does not really understand a given body of material until he or she can communicate that material to someone else. This suggests that in proceeding through this book the student should attempt to master the material to the point where it could be taught to a colleague.

I have often thought that in ideal circumstances it would be best to require one's students to teach statistics to their roommates and that the tests should be administered to the roommates rather than to the students themselves. Should one do so, there would be no doubt that when the roommates passed the test their instructors would have mastered the material.

Even though it is impractical to examine students in this fashion, it is still advisable for them to pursue the topic under discussion to the point where they could, if need be, teach it. This, of course, places a heavy responsibility on the individual for it requires a constant monitoring of one's mastery. There can be little doubt that one of the most difficult tasks that any of us face is recognizing when our thoughts on a given topic are fuzzy and incomplete. Still, the gains from doing so are certain to be worth the effort. After all, the development of statistical reasoning is one of the great intellectual accomplishments of this century, and whether one is aware of it or not, statistical matters permeate every aspect of modern thought. Certainly no member of this generation can be considered to be fully educated if they have failed to acquire the powerful analytical tools that are provided by a first class knowledge of statistical reasoning. This book and its computer exercises are designed to enable the student to master those tools. There are eight such exercises. Each covers a different topic, and each provides an opportunity to explore its topic in a direct empirical fashion.

The material in Appendix II includes descriptions of each exercise along with instructions for its use. The Table of Contents provides the titles of these exercises, and it indicates which exercise is intended to accompany each chapter.

A Note to the Instructor

Extensive use of the computer exercises can be especially useful in elaborating the concepts covered in this book. Most of these exercises take only a minute or two to run yet each provides a full printout that can be studied later. For these reasons, access to only one or two computers (and printers) should be sufficient for even a reasonably large class.

Each of the computer exercises is designed to enable students to explore a full range of the factors that contribute to the concepts it elaborates. I have found, for example, that if used properly, the computer exercises on the sampling distribution of $z_{\bar{x}}$ and t can provide considerable insight into the concept of statistical power. This is often among the most difficult of ideas to grasp, but if students are encouraged to use these exercises to explore the effects of various sample sizes when an hypothesis they elect to test is false by various amounts as well as when it is true, they can discover for themselves the subtle interrelationships that exist between α, β, power and n.

Finally, I must mention that other than the computer exercises, and the numerical examples included in the text, this work contains no problems or study exercises. My own preference is to use real data for such exercises. I can usually acquire such data from my own research or from the work of my colleagues. Also, I have found that there are numerous statistical workbooks available if the kind of problems and drills they provide seem likely to be useful.

Table of Contents

Chapter 1

STATISTICAL APPROACH

In a very real sense, most of the decisions we make in our life are based on incomplete data and hence, properly speaking, are statistical decisions. We we choose to attend a given college or university we may examine various catalogues and we may even elect to visit several schools, but in the end we must make our decision well before we have acquired all the pertinent data. Indeed, even when we elect to cross a dangerous blind intersection we are exemplifying the statistical decision that the odds of our being injured are low enough to be discounted. The process is generally quite straighforward. We collect as much information as seems reasonable (We study the college catalogues, we listen for possible approaching vehicles) and then we come to a decision based on our best guess or hunch as to its implications.

In the sciences, one must also frequently make decisions based on what are incomplete sets of data. In such instances, however, instead of merely relying on hunches we try to carefully reason through all the available information before coming to an opinion.

Often, as might be expected, we find that even in the sciences our initial hunches about a given body of data are very close to the opinion we exhibit after we have employed a systematic analytic approach. But sometimes we encounter surprises. Here is an interesting example that illustrates the surprising results that can be generated when one proceeds systematically through a given problem.

Imagine that the earth is a perfectly smooth sphere and that you have somehow managed to pass a wire around its circumference so that the fit is snug at every point. Now imagine that you add 100 feet to that wire and that you are faced with the task of taking up the slack by raising the wire onto poles that will be erected all the way around the circumference of the earth. The problem before you is to determine how tall the poles must be.

When faced with this question most students recognize that 100 feet is a very small distance to add to such a long wire and accordingly they assume that the poles will have to be very, very small, perhaps a thousandth of an inch tall, maybe even less. Now let's examine the problem using a systematic analytical approach.

Let C = the circumference of the earth in feet.

r = the radius of the earth in feet.

π = the mathematical constant 3.14

For any circle $C = 2\pi r$.

Our problem asks how much one must increase the radius of a circle if the circumference of the circle is increased by a given amount. (i.e., 100 ft)

Stating our problem algebraically we get

$$(C + 100) = 2\pi(r + X)$$

but $C = 2\pi r$ therefore

$$(2\pi r + 100) = (2\pi r + 2\pi X)$$

where X is the increase in the radius

If we subtract $2\pi r$ from both sides we get

$$100 = 2\pi r$$

dividing both sides of the equation by 2 yields

$$\frac{100}{2\pi} = X$$

i.e., X = 15.9 feet

Apparently, if we are to take up 100 feet of slack in the wire, the poles would have to be almost 16 feet tall. This unexpected result is perfectly sound and it provides us with the interesting proposition that if you add a quantity to the circumference of a circle the radius will be increased by a constant amount regardless of what the circumference was. In other words the same answer would have been obtained if it were suggested that you add 100 feet to a wire that passed around a basketball or 100 feet of wire that passed around our galaxy. In both cases the increase in the radius of the circle would be approximately 16 feet.

Statistical reasoning is much like the example just cited in that it provides us with a way of being quite explicit about the procedures we use to draw inferences. More specifically, it gives us techniques for visualizing and describing a given set of data. It provides us with an exact methodology for making decisions based on the data, and perhaps most importantly, it provides us with a way of specifying the degree of confidence we have in our decision.

Techniques for Visualizing Data

Histograms

Listed below is a set of test scores from a small class in English literature. Our first task will be to find some convenient way of visually representing this distribution of scores.

Numerical Scores

92.03	69.12	68.14	87.18
74.32	75.41	70.23	87.16
91.88	68.01	82.31	73.27
70.01	54.88	68.99	81.46

As we study these data we almost immediately note that no two items are exactly the same. This should come as no surprise since the instructor has calculated the test scores to two decimal places, thereby allowing for rather fine distinctions. Of course, whether or not such fine distinctions are really meaningful is open to debate. For the moment, however, we should note that a faithful visual representation of the data (as given) would exhibit the rather awkward form seen below.

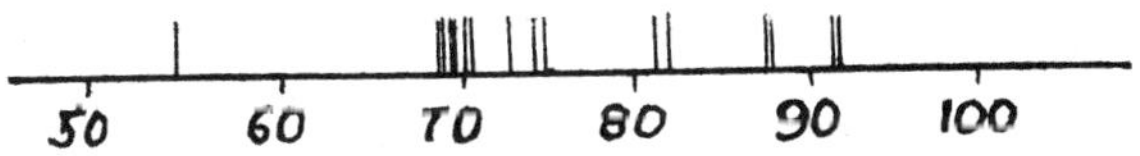

While this figure accurately depicts all of the information in the raw data one cannot readily extract its essential features. What seems needed is to somehow arrange the scores in a format that emphasizes the major features of the data while perhaps sacrificing some of the details. One way this can be done is to plot the scores in terms of some predetermined set of class intervals. For example, the usual grading system runs as follows:

A = 90 - 100
B = 80 - 99
C = 70 - 79
D = 60 - 69
F = -- - 59

Unfortunately, if we attempted to plot our data according to the above class intervals we would be unable to represent the score 69.12 because it falls between two of the class intervals. This problem could have been avoided if we had been more judicious in our specification of intervals.

Let's try again, only this time we will establish class intervals that will leave no room for ambiguity. For that purpose we'll resort to the following strategem.

A = 90 - 100.
B = 80 - $89.\dot{9}$
C = 70 - $79.\dot{9}$
D = 60 - $69.\dot{9}$
F = -- - $59.\dot{9}$

In this array, the quantity $69.\dot{9}$ is understood to represent 69.9999999999 etc. That is the dot over the 9 means that the 9 is repeated indefinitely. Clearly, with this arrangement every score will fall into one or another class interval.

Here is the distribution that emerges when we plot the scores in this fashion.

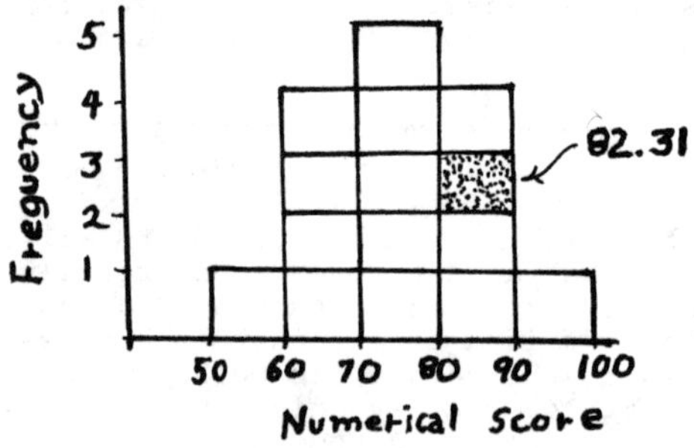

A distribution of this sort is called a histogram. In a histogram each score is represented by a area. For example, the score 82.31 is represented by the indicated area. The histogram provides a view of the data that seems to convey its basic structure in terms that are readily comprehended by the observer. Of course, we have sacrificed some precision in order to arrive at this representation, but when one is interested in viewing the underlying structure in a set of data, such sacrifices are often necessary.

By now the thoughtful student will have recognized that there is a certain degree of arbitrariness that goes into the formation of any histogram and that the shape of the histogram will, in part, reflect the decisions one makes with respect to the nature of the class intervals. Indeed, had only slightly different class intervals been used in the above example, the distribution would have had a rather different shape. What can one say about such arbitrary procedures? Not much, except to point out that there simply are no firm rules about these matters and that how one portrays a given distribution is often largely determined by one's purposes.

Frequency Polygons

Here is another way to visually represent the distribution of test scores presented earlier.

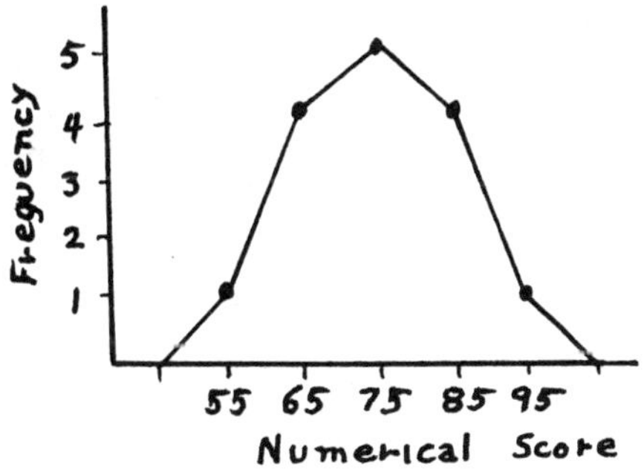

As in a histogram, each score in a frequency polygon is conceived to occupy a given area under the curve. In general, the formation of a frequency polygon requires that one first arrange the raw scores according to a predetermined set of intervals. Ordinarily these would be the same class intervals that would be used if we had planned to plot the data in histogram form. In forming a frequency polygon, however, we erect ordinates at the midpoints of the class intervals and then connect them to each other.

If properly constructed a frequency polygon and histogram will have equal areas. How this comes about can be seen below.

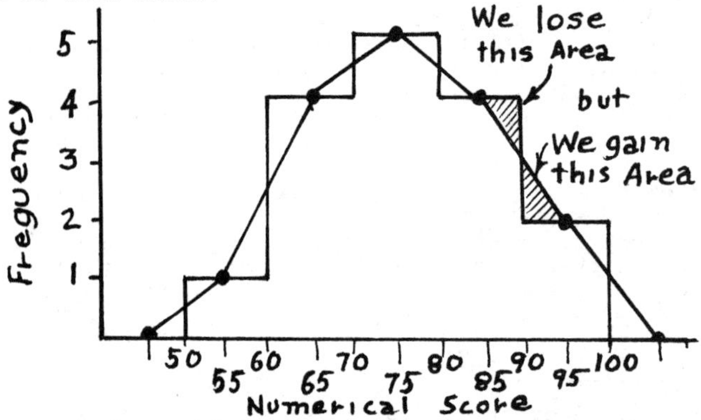

Here is a third way of visually representing the data that we have been dealing with. The function is called a cumulative distribution and it can take either of the two forms seen below.

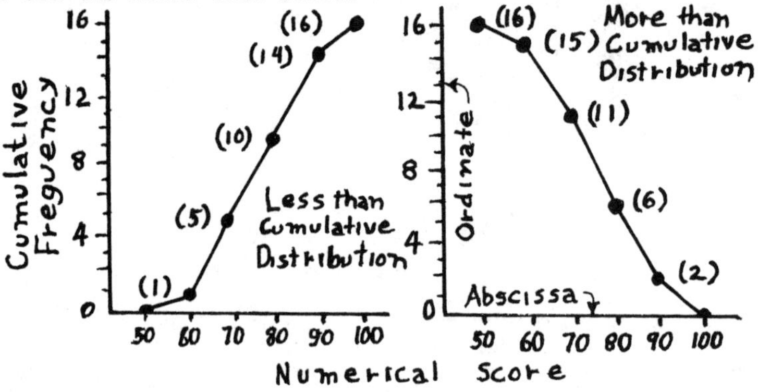

The ordinate (the vertical dimension) of the function at the right shows the number of scores in the distribution that are equal to or greater than the value indicated on the abscissa (the horizontal dimension) of the function. The ordinate of the cumulative distribution on the left shows the number of scores in the distribution that are equal to or less than each value indicated on the abscissa. Often the ordinate of a cumulative distribution will be expressed in percentages. Representing a distribution in this fashion permits one to easily visualize the points along the scale of measures that cut off various percentages of the scores.

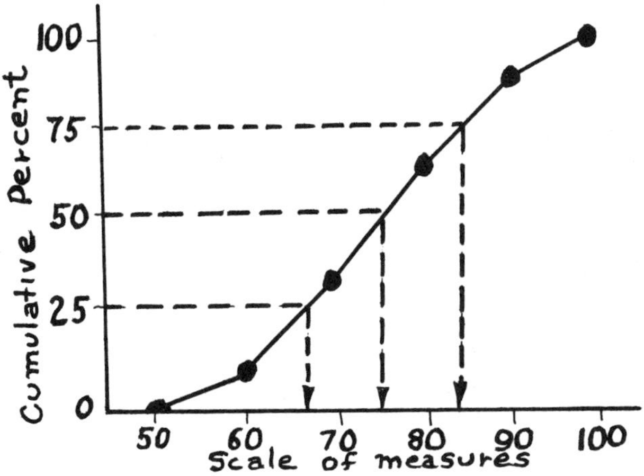

Describing a Given Distribution Using Measures of Central Tendency and Dispurson

As indicated by our earlier discussion, a given distribution must fall somewhere along the scale of measures. For example, the next page shows two histograms plotted by a farmer with some sophistication about statistics. The histogram to the left indicates weights of lambs, whereas the histogram to the right represents weights of sheep.

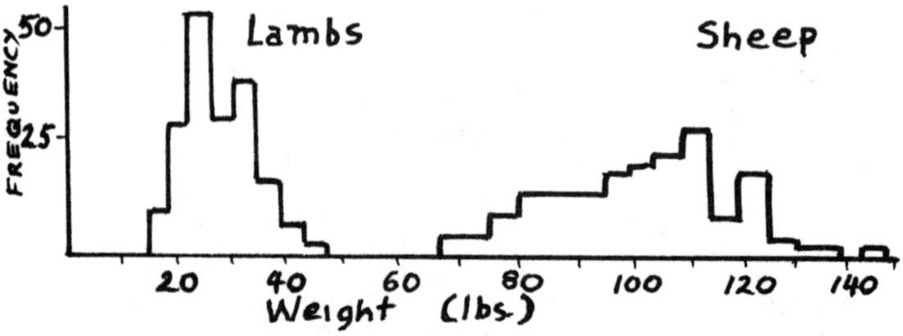

The point along the scale where the distribution of weights of lambs seems to center is about 30 pounds. The point where the distribution of weights of sheep centers is about 110 pounds. We can also see that the two distributions occupy different amounts of the scale.

If we can indicate the point (along the scale) where the center of a distribution falls and if we can also indicate the distance (along the scale) occupied by the distribution we will have gone a long way toward summarizing the information in that distribution.

Measures of Central Tendency

(The median)

One way to specify the point where the center of a distribution falls is to indicate the point that divides the distribution into two equal parts. This point is called the median.

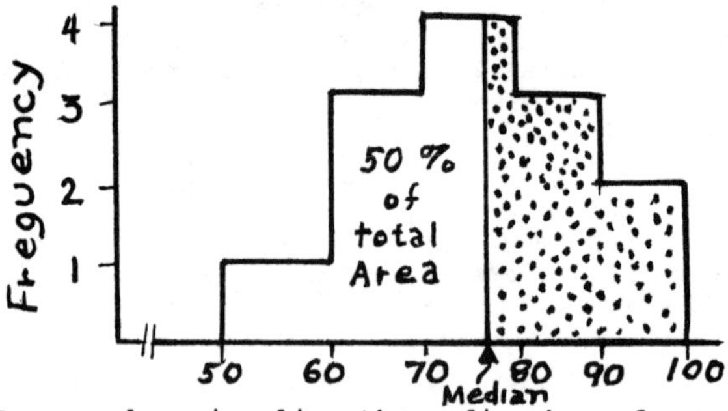

We can also visualize the median by reference to a cumulative distribution expressed in percentiles (i.e., percentages).

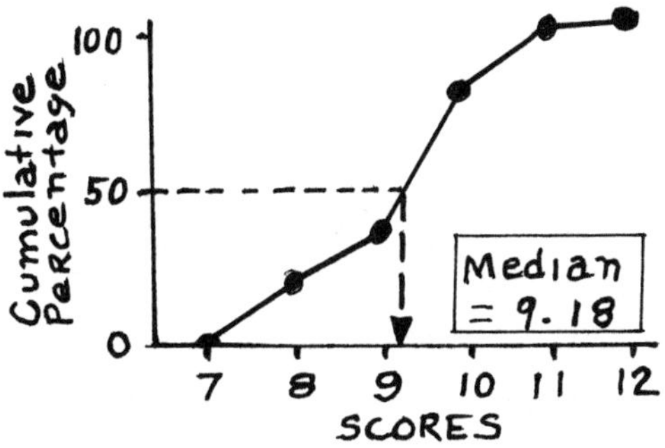

(The mode)

A second way to indicate the location of a given distribution is to specify the point along the scale of measures where the most frequent score occurs. That point is called the mode of the distribution. In this regard it is important to emphasize that the mode of a given distribution is a point along the scale of measures. It is not the frequency of the most common score.

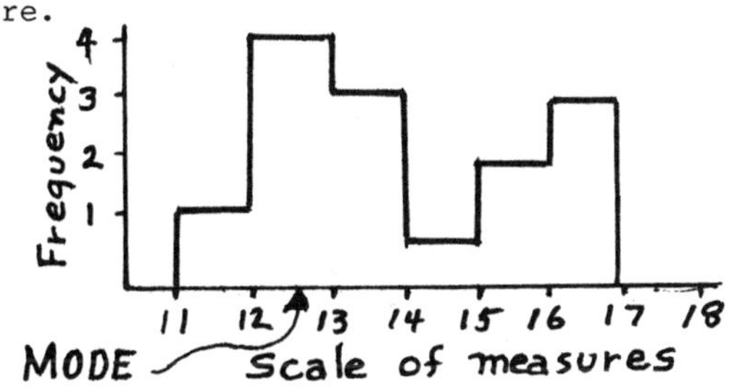

(The mean)

A third and especially useful way to specify the point where a distribution centers is to calculate the average of the scores in the distribution. As you surely know, the average of a set of scores is their sum divided by the number of cases. For example, the average of 1, 2, 2, 3, 3, 3, 4, 4, 5 is 3 because

$$\frac{1 + 2 + 2 + 3 + 3 + 3 + 4 + 4 + 5}{9} = \frac{27}{9} = 3$$

In other words, to compute the average (or mean)

of a set of numbers, one must carry out a specific sequence of arithmetic calculations. Statisticians are accustomed to using certain algebraic symbols to describe such sequences of operations.

For example they use the expression $\frac{\sum_{1}^{N} X}{N}$ to describe the operations used to calculate the mean. In this expression the Greek letter $\sum$ is understood to be an instruction to carry out the operation of adding a number of Xs. More specifically the symbol $\sum_{1}^{N} X$ tells us to add together all the Xs from the 1st to the Nth.

Here is a set of six Xs

$X_1 = 3 \quad X_2 = 7 \quad X_3 = 9$

$X_4 = 11 \quad X_5 = 11 \quad X_6 = 12$

For these Xs the expression

$$\sum_{1}^{N} X = X_1 + X_2 + X_3 + X_4 + X_5 + X_6$$
$$= 3 + 7 + 9 + 11 + 11 + 12$$
$$= 53$$

and the expression

$$\frac{\sum_{1}^{N} X}{N} = \frac{53}{6} = 8.83$$

There are times when it is necessary to instruct one to only add up a restricted portion of the total number of Xs. For example, the expression $\sum_{2}^{5} X$ tells us to add only those Xs with subscripts = 2, 3, 4, and 5.

thus $$\sum_{2}^{5} X = 7 + 9 + 11 + 11 = 38$$

Most of the time, however, it will not be necessary to specify the limits of the summation because we will be summing across all N items before us. Thus, the mean of the distribution of the 16 original test scores we discussed earlier is specified as follows:

$$\frac{\sum X}{N} = \frac{1214.40}{16} = \boxed{75.90}$$

It is of interest that the mean of the raw test scores obtained here is quite similar to the mean of the scores when they have been arranged in a histogram. We saw earlier that in forming a histogram to represent a given set of scores we must begin by specifying the exact boundaries of the several class intervals. Having done so, we then consider a given score as being represented by an area within the appropriate class interval. We can calculate a mean on data that have been arranged into class intervals if we take the midpoint of a given class interval to represent each of the scores in that interval. If we do so for the histogram of the set of test scores discussed earlier we find that

$$\frac{\sum X}{N} = \frac{1}{16}\left(55 + 65 + 65 + 65 + 65 + 75 + 75 + 75 + 75 + 75 + 85 + 85 + 85 + 85 + 95 + 95\right)$$
$$= \frac{1220}{16} = \boxed{76.25}$$

which is reasonably close to the value 75.90 that we calculated for the original data.

Arithmetically, the mean of a distribution is that score which could be substituted for every score in the distribution and still yield the same total.

This can be easily shown as follows:

1) By definition Mean $= \frac{\sum X}{N}$

Now the equal sign indicates that the quantity on the left side of the equation is exactly equivalent to the quantity on the right side. This in turn implies that if we change the quantity on one side of the equation we must change the other side by the same amount. For example if we multiply the left side of the above equation by N we must also multiply the other side by N.

$$\text{i.e., } N(\text{Mean}) = N\frac{\Sigma X}{N} \quad \text{but } N\frac{\Sigma X}{N} = \frac{N}{N}\Sigma X = \Sigma X$$

In other words if we substitute the mean for each of the Xs in a distribution we wind up with N means and their sum will be the same as the sum of the original Xs.

Another property of the mean is that for a given distribution, the sum of the deviations from the mean is always zero. To see why this is so we will, as before, begin by considering a set of Xs

$$X_1, X_2, X_3, \cdots X_N$$

Statisticians often use the Greek letter μ (mu) to represent the mean of such a set of Xs.

Algebraically $$\mu_X = \frac{\Sigma X}{N}$$

If we subtract μ_X from each item in the set we will obtain a new set of items each of which represents the deviation of a given score from the mean.

Here is that set of deviations.

$$(X_1 - \mu_X), (X_2 - \mu_X) \cdots (X_N - \mu_X)$$

To express the sum of these deviations we need merely write the expression on the next page.

$$\Sigma(X-\mu_X) = (X_1-\mu_X)+(X_2-\mu_X)+\cdots+(X_N-\mu_X)$$

$$= \Sigma X - N\mu_X$$

$$= \Sigma X - N\frac{\Sigma X}{N}$$

$$= \Sigma X - \Sigma X$$

$$= 0$$

Note that there are N μ_Xs because there are N Xs

One consequence of this property of the mean is that the mean of a distribution can be conceived to represent a kind of balance point for the distribution. That is, if for some reason we constructed our distribution out of blocks of wood, the mean would be the point where the assembly would balance.

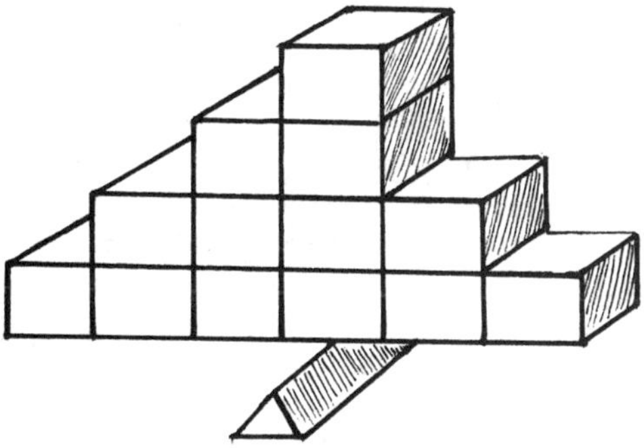

Although either the mode, median or mean can tell us something about the point along the scale of measures where the distribution is located, none of these measures provides any information about the variability of the scores as reflected in the amount of scale covered by the distribution.

Measures of Variability

(The range)

Perhaps the most obvious way to specify the variability of a given distribution is to indicate the distance (along the scale of measures) between the highest and lowest scores. This index-of-variability is called the range.

Algebraically

$$\text{range} = \text{highest score} - \text{lowest score}$$

Although we might think at first glance that the range would be an ideal measure of variability this is not the case. The problem is that the range is completely dependent upon the two most extreme scores and it completely ignores all of the other scores in the distribution. As a result the two distributions below exhibit the same range but except for the two extreme values, the scores are much more variable in the distribution on the left.

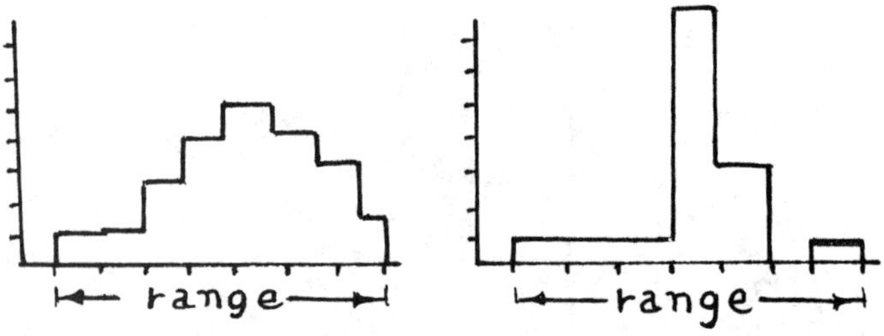

(The interquartile Range)

A somewhat more sophisticated index of the variability in a distribution is the distance between the first and third quartiles, i.e., the interquartile range. As the name implies the quartiles of a distribution are the points along the scale of measures that divide the distribution into quarters,

i.e., the first quartile (Q_1) cuts off the lower 25%

of the distribution.

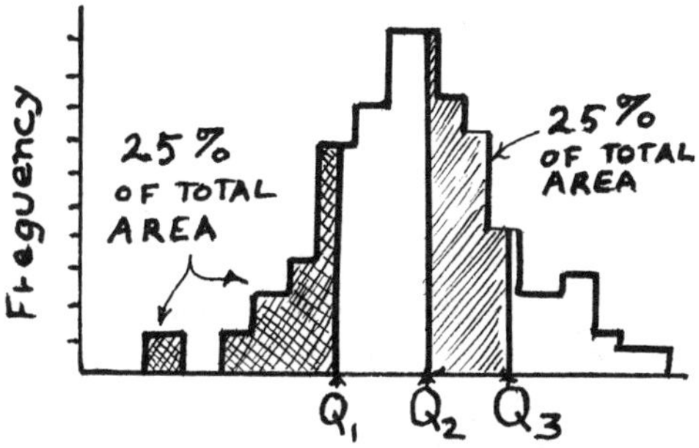

Note that Q_2 is the MEDIAN

Clearly the interquartile range will be large when the scores in a given distribution are variable, and it will be small when the scores are generally close to one another. A variation on the interquartile range that statisticians sometimes employ is the semi-interquartile range.

Algebraically the

Semi-interquartile range
= 1/2 (the interquartile range)

(The Average Deviation)

Another measure of variability in a given set of scores is the average of the absolute deviations from the mean.

We have previously seen that in a given distribution: $\sum(X - \mu_X) = 0$. This implies that in any distribution the $\frac{\sum(X - \mu_X)}{N}$ will also equal zero. The average deviation has a format that is similar but not identical to this equation.

Algebraically the average deviation is specified

as follows:

$$\text{Average deviation} = \frac{\Sigma|X - \mu_X|}{N}$$

In this equation the expression $|X - \mu_X|$ instructs us to calculate the distance between each score and the mean and then to specify that distance without regard to whether the score is above or below the mean. In short we are told to calculate the absolute value of each deviation from the mean and the average deviation is defined as the mean of these absolute values.

X	$(X - \mu_X)$	$\lvert X - \mu_X \rvert$
$X_1 = 1$	$(1-3) = -2$	$\lvert 1-3 \rvert = 2$
$X_2 = 2$	$(2-3) = -1$	$\lvert 2-3 \rvert = 1$
$X_3 = 2$	$(2-3) = -1$	$\lvert 2-3 \rvert = 1$
$X_4 = 3$	$(3-3) = 0$	$\lvert 3-3 \rvert = 0$
$X_5 = 3$	$(3-3) = 0$	$\lvert 3-3 \rvert = 0$
$X_6 = 3$	$(3-3) = 0$	$\lvert 3-3 \rvert = 0$
$X_7 = 4$	$(4-3) = +1$	$\lvert 4-3 \rvert = 1$
$X_8 = 4$	$(4-3) = +1$	$\lvert 4-3 \rvert = 1$
$X_9 = 5$	$(5-3) = +2$	$\lvert 5-3 \rvert = 2$
$\Sigma X = 27$	$\Sigma(X - \mu_X) = 0$	$\Sigma \lvert X - \mu_X \rvert = 8$

$$\mu_X = \frac{\Sigma X}{N} = \frac{27}{9} = 3$$

$$\text{Average Deviation} = \frac{\Sigma|X - \mu_X|}{N} = \frac{8}{9} = 0.888$$

(The Variance)

Although all of the above measures of variability have their uses, we have dealt with them primarily to set the stage for a discussion of another measure of variability known as variance. By definition the variance is equal to the mean of the squared deviations from the mean.

Statisticians usually use the Greek letter σ^2 to represent the variance. More specifically

$$\sigma_X^2 = \frac{\Sigma(X - \mu_X)^2}{N}$$

Once again the equation can be conceived to contain a set of instructions. They tell us to calculate the mean $\mu_X = \frac{\Sigma X}{N}$ of the set of X_s. Then we must subtract μ_X from each X and square the resulting quantity. The variance is the average of those squared quantities.

For example, given the following set of

Values of X	$(X-\mu_X)$	$(X-\mu_X)^2$
1	(1-5)=-4	16
3	(3-5)=-2	4
3	(3-5)=-2	4
5	(5-5)= 0	0
5	(5-5)= 0	0
5	(5-5)= 0	0
5	(5-5)= 0	0
5	(5-5)= 0	0
7	(7-5)=+2	4
7	(7-5)=+2	4
9	(9-5)=+4	16
ΣX=55	$\Sigma(X - \mu_X)$=0	$\Sigma(X-\mu_X)^2$=48

$$\mu_X = \frac{\Sigma X}{N} = \frac{55}{11} = 5 \qquad \sigma_X^2 = \frac{\Sigma(X-\mu_X)^2}{N} = \frac{48}{11}$$

$$= 4.36$$

While it is clear that the variance of a distribution will be large when the scores are spread out and it will be small when the scores are close together, the variance does not itself represent a distance along the scale of measures. Instead, it is an average of a set of squared distances which could perhaps be interpreted as some sort of area. We can, however, always convert a given variance into a measure that expresses distance by taking its square root. The resulting quantity σ_X is called "the standard deviation".

Algebraically $\sigma_X = \sqrt{\sigma_X^2} = \sqrt{\frac{\sum(X-\mu_X)^2}{N}}$

For the distribution of test scores in English literature that we studied earlier, the variance calculated on the original scores is numerically equal to 92.3487. Accordingly, the σ_X of that set of scores must be 9.6098.

Suppose, however, that we have reason to calculate the variance on the test scores after they have been arranged in a histogram. To do so, we would use the midpoints of the appropriate class intervals

to represent each item. The resulting σ_X^2 would be calculated as follows:

$$\frac{\sum(X-\mu_X)^2}{N} = \frac{1}{16}\left[(55-76.25)^2+(65-76.25)^2+ \cdots + (95-76.25)^2\right]$$

$$= \frac{1}{16}\left[(451.56)+(126.56)+ \cdots + (351.56)\right]$$

$$= \boxed{115.72}$$

Note that here we use $\mu_X = 76.25$ which we saw earlier is the mean of the data when they have been arranged in a histogram.

As was the case when we compared the mean calculated on raw scores vs. grouped scores, again we find relatively close agreement. As to which is the more accurate, or better way to summarize the data, it is largely a matter of expediency. If one had need to display the histogram it would be reasonable to calculate the mean and variance of the grouped data as above. On the other hand, if there were no need to display the data one would probably want to calculate the mean and variance on the raw scores. As is usual in most matters of human endeavor, the choice is determined largely by one's purposes.

The Effects of Transforming the Values of X

Here is a distribution of X_s with $\mu_X = 7$ and $\sigma_X = 1.8$

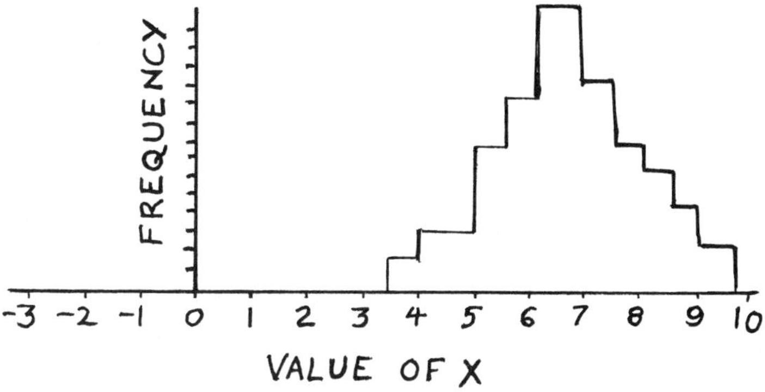

We can move the distribution up or down the scale of measures by adding or subtracting a constant from each item (i.e., each X) in the distribution.

For example, if we subtract 5 from every X in the above distribution we move the distribution down the the scale by 5 units, but we do not change its variability.

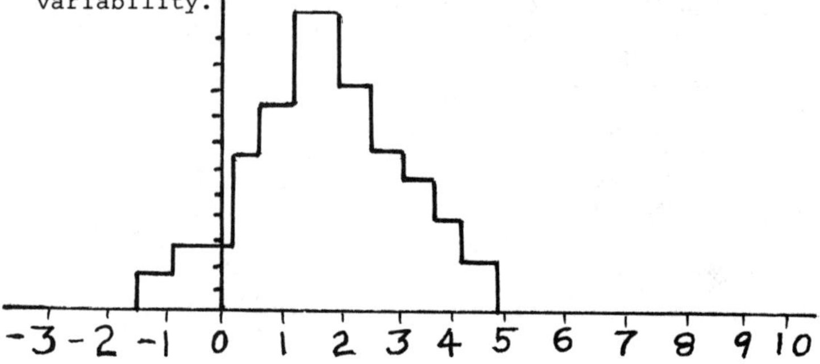

In general, if we start with a set of X's with

mean = μ_X and variance = σ_X^2 and we add a constant (C) to each X we get a new distribution

$$(X_1 + C), (X_2 + C), \cdots (X_N + C)$$

The mean of this new distribution will equal $\mu_X + C$ C because

$$\frac{\Sigma(X+C)}{N} = \frac{\Sigma X + NC}{N} = \frac{\Sigma X}{N} + \frac{NC}{N} = \mu_X + C$$

The variance of the new distribution will, however, be unchanged because

$$\sigma^2_{(X+C)} = \frac{\Sigma[(X+C) - (\mu_X + C)]^2}{N} = \frac{\Sigma[X + C - \mu_X - C]^2}{N}$$

$$= \frac{\Sigma(X - \mu_X)^2}{N} = \sigma_X^2$$

Note that we just showed that (μ_X + C) is the mean of the distribution of (X + C)s

If we multiply every item in a distribution by a constant (c) we create a new distribution

$$CX_1, \; CX_2, \; CX_3 \cdots CX_N$$

with $\mu_{CX} = C\mu_X$

$$\sigma^2_{CX} = C^2\sigma_X^2$$

$$\sigma_{CX} = C\sigma_X$$

This comes about as follows:

$$\mu_{CX} = \frac{\sum CX}{N} = \frac{C\sum X}{N} = C\mu_X$$

$$\sigma^2_{CX} = \frac{\sum (CX - C\mu_X)^2}{N}$$

This (as we just saw) is the mean of the CXs

$$= \frac{\sum C^2(X - \mu_X)^2}{N}$$

When we move C outside the parenthesis we must square it because the parenthesis is squared

$$= \frac{C^2\sum (X - \mu_X)^2}{N}$$

$$= C^2\sigma_X^2$$

$$\sigma_{CX} = \sqrt{\sigma^2_{CX}} = \sqrt{C^2\sigma_X^2} = C\sigma_X$$

The thoughtful student will recognize that by specifying the effects of adding a constant and multiplying by a constant we also specify the effects of subtracting a constant and dividing by a constant. This follows because subtracting a constant is the equivalent of adding a negative quantity and dividing by a constant is the equivalent of multiplying by a fractional quantity. In the material which follows we will have many occasions to transform a given distribution of scores into a new distribution by either adding a constant to the value of each item in the distribution or by multiplying the value of each item in the distribution by a constant.

This means that it will be quite important to acquire a thorough grasp of the transformation effects we have just elaborated. For this purpose some students may find it useful to summarize those effects in words.

> In general: Adding a constant to every item in a distribution adds the constant to the mean of the distribution, but it leaves the variance of the distribution unchanged.
>
> Multiplying every item in a distribution by a constant multiplies the mean and standard deviation of that distribution by the constant and it multiplies the variance of the distribution by the square of the constant.

<u>The Z Score Conversion</u>

<u>A special case of a uniform transformation of X</u>

Here is a distribution of X's with μ_X = 41.3 and σ_X = 7.24

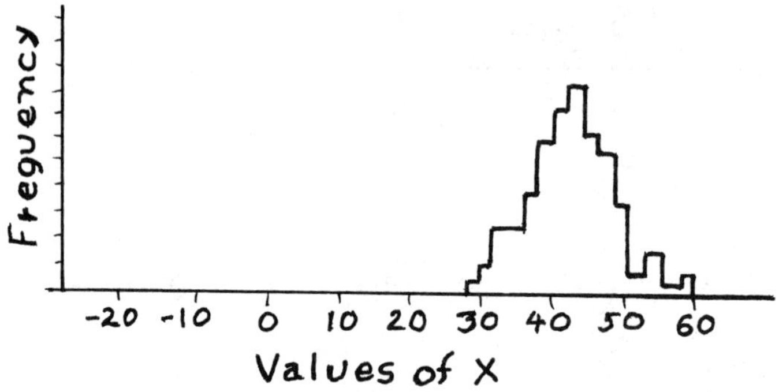

If we subtract 41.3 (i.e., the mean of the distribution) from every item in the original distribution we get a new distribution with

$$\mu_{(X-\mu_X)} = 0$$

$$\sigma_{(X-\mu_X)} = 7.24 = \sigma_X$$

Here is that distribution

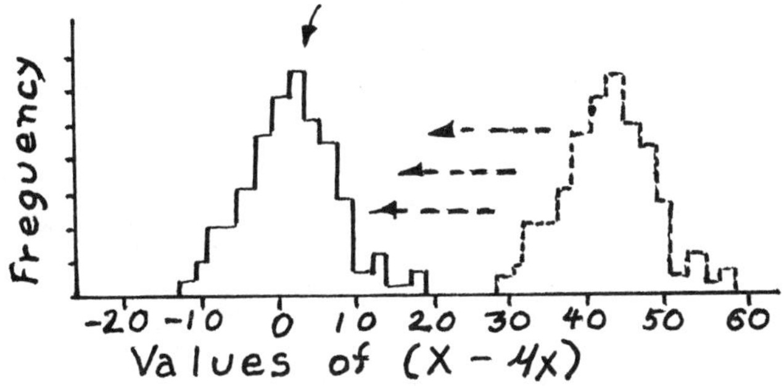

If we now divide every item in this new distribution by 7.24 (i.e., by σ_X) we get a third distribution with mean = 0 and variance = 1.

This follows because we divide the mean by 7.24 but 0/7.24 = 0. We divide the standard deviation by 7.24 but 7.24/7.24 = 1.

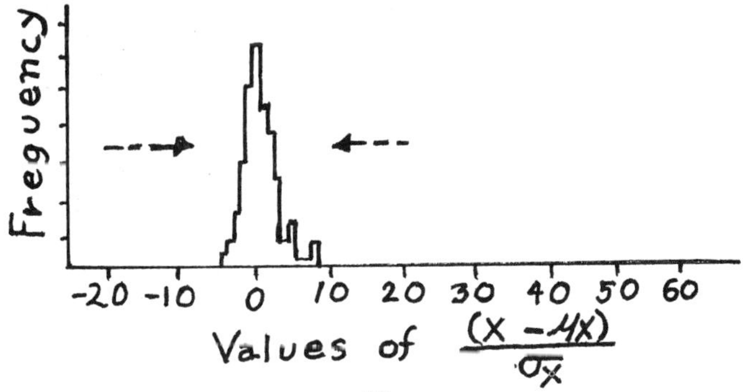

In this distribution

$$\mu\left(\frac{X-\mu_X}{\sigma_X}\right) = 0 \qquad \sigma\left(\frac{X-\mu_X}{\sigma_X}\right) = 1$$

This leads us to the important general proposition that any distribution of X's can be transformed into a new distribution with mean = 0 and σ = 1 by converting each X into what statisticians call a Z score.

By definition $Z_X = \dfrac{X-\mu_X}{\sigma_X}$

Whenever a distribution of Xs is converted to Z scores μ_Z will equal 0 and σ_Z will equal 1

Here is the formal proof of this proposition:

$$\mu_Z = \frac{\Sigma Z}{N} = \frac{1}{N}\sum\left(\frac{X-\mu_X}{\sigma_X}\right) = \frac{1}{N\sigma_X}\sum(X-\mu_X)$$

but as we saw earlier

$$\sum(X-\mu_X) = 0 \quad \text{thus } \frac{1}{N\sigma_X}[0] = 0$$

$$\sigma_Z^2 = \frac{\sum(Z-\mu_Z)^2}{N} \quad \text{but } \mu_Z = 0$$

$$\text{thus} \quad \sigma_Z^2 = \frac{\Sigma Z^2}{N} = \frac{\Sigma\left(\frac{X-\mu_X}{\sigma_X}\right)^2}{N}$$

$$= \frac{1}{N\sigma_X^2}\sum(X-\mu_X)^2 = \frac{1}{\sigma_X^2}\left(\frac{\Sigma(X-\mu_X)^2}{N}\right)$$

$$= \frac{1}{\sigma_X^2}\left(\sigma_X^2\right) = 1$$

The X and Z transformation is especially useful when one wishes to compare scores across two or more distributions.

For example, a student gets an 87 in an English exam where the mean was 83.1 and the standard deviation was 4.2. The same student gets an 81 in a History exam where the mean was 78 and the standard deviation was 3. We can evaluate the student's relative performances on the two exams by using a score transformation.

English		History	
Raw exam score	X = 87	Raw exam score	X = 81
μ_X for the class	= 83.1	μ_X for the class	= 78
σ_X for the class	= 4.2	σ_X for the class	= 3

$$Z_X = \frac{87 - 83.1}{4.2} = .92 \qquad Z_X = \frac{81 - 78}{3} = 1$$

These calculations tell us that the score in history was one standard deviation above the mean of all history students. The score in English, on the other hand was only .92 standard deviations above the mean of the students in the English class. Despite the differences in the raw scores, the student did somewhat better in history than in English.

Chapter 2

THE NORMAL CURVE

When one measures a very large number of items and arranges to graphically represent the data using very fine class intervals the histogram that results is often somewhat bell shaped.

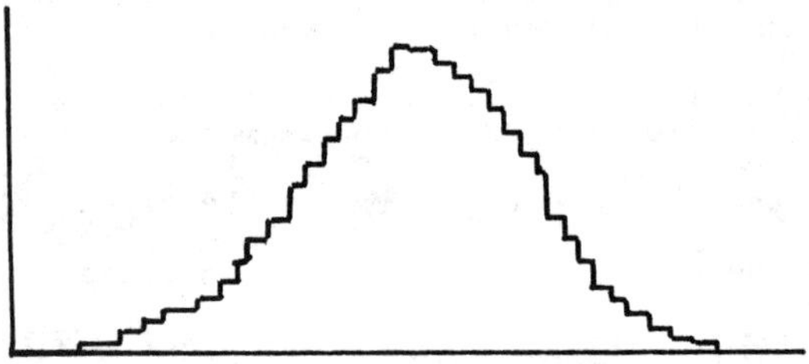

Statisticians describe distributions of this sort as being normal when the underlying function is specified by the following equation:

$$y = N i \left[\frac{1}{\sigma_X \sqrt{2\pi}} \right] e^{-\frac{1}{2}\left(\frac{X - \mu_X}{\sigma_X}\right)^2}$$

In order to grasp the important message conveyed by this rather complex equation it will be helpful if we first review some basic notions about equations.

Here, for example, is a very simple equation:

$$Y = .5\,X$$

When mathematicians encounter this equation they understand it to mean that exactly what it looks like it says, namely that the value of Y is always 1/2 the value of X.

Thus if X = 10 Y must equal 5
and if X = 4 Y must equal 2.

Here is a table representing an arbitrarily selected set of values of X along with the values of X that the equation specifies.

X	Y
-2	-1
0	0
+3	1.5
+5	2.5

As you can see, the table reveals nothing more than the basic idea specified in the equation itself, namely that

$$Y = .5X$$

Equations of this sort can be represented graphically if we plot several of its points in a set of Cartesian Coordinates.

For example here is a plot of the data in the above table.

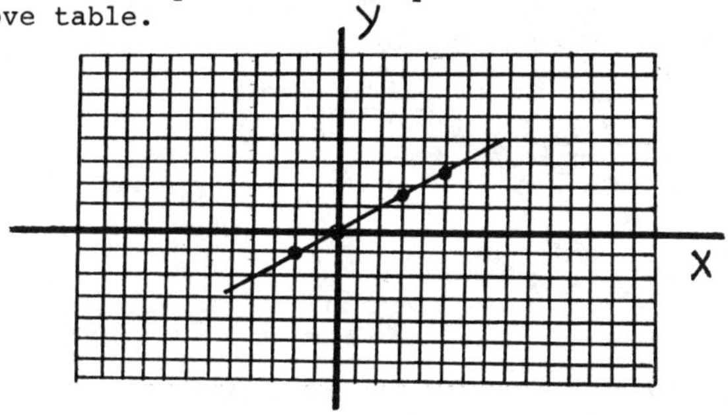

As can be seen, the points fall along a straight line that passes through the origin.

Here is the line specified by the equation $Y = 2X + 3$

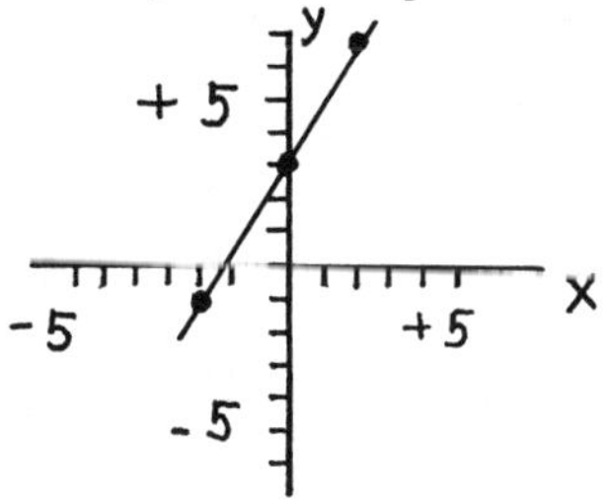

X	Y
0	3
-2	-1
2	7

This time the line has greater slope and it does not pass through the origin

Both of the equations we have just examined are special cases of the more general equation for a straight line.

$$Y = bX + a$$

An equation of this sort is said to contain two parameters (b & a). As we have just seen, the value of b determines the slope of the line whereas the value of a specifies the Y intercept (i.e., the value of Y when X = 0).

Note: In the equation Y = .5X the value of a = 0

Now let's re-examine the equation that specifies a normal curve. Suppose, for example, we wanted to fit a normal curve to a given histogram, that is, suppose we wanted to plot that unique normal curve that has the same area, σ_X and μ_X as a given histogram. To do so we would use the following equation:

$$Y = Ni\left[\frac{1}{\sigma_X\sqrt{2\pi}}\right] e^{-\frac{1}{2}\left(\frac{X - \mu_X}{\sigma_X}\right)^2}$$

In this equation

X = a value along the abscissa
Y = the height of the ordinate at a given value of X
N = the number of cases in the histogram
i = the length of a class interval in the histogram (note (N)(i) = the total area in the histogram)
σ_x = the standard deviation of the histogram
μ_x = the mean of the histogram
π = a mathematical constant = 3.1416
e = a mathematical constant 2.7183

To help keep our computations in order it will be helpful to rearrange the equation as follows:

In the earlier form of this equation we had a negative exponent. In high school algebra we learned that negative exponent is an instruction to take a reciprocal.

$$10^{-3} = \frac{1}{10^3} = \frac{1}{(10)(10)(10)} = \frac{1}{1000}$$

Suppose we want to plot a normal curve with the same area mean and variance as a histogram with 100 items, class interval = 1 μ_x = 37 σ_x = 4. If we now replace the various parameters and constants in the above equation by the appropriate numbers we get

$$Y = \left[\frac{(100)(1)}{4\sqrt{2\,(3.1416)}}\right] \times \left[\frac{1}{2.7183^{\frac{1}{2}\left(\frac{X-37}{4}\right)^2}}\right]$$

therefore

$$Y = 9.9735 \left[\frac{1}{2.7183^{\frac{1}{2}\left(\frac{X-37}{4}\right)^2}}\right]$$

This then is the equation that specifies a normal curve with total area = 100 units and with μ_x = 37 and σ_x = 4.

Our next step would be to calculate the value of Y for various arbitrarily selected values of X. Here are those calculations. The column headings indicate the steps in the calculations.

X	$\frac{X-37}{4}$	$\frac{1}{2}\left(\frac{X-37}{4}\right)^2$	$2.7183^{\frac{1}{2}\left(\frac{X-37}{4}\right)^2}$	$\frac{1}{2.7183^{\frac{1}{2}\left(\frac{X-37}{4}\right)^2}}$	$9.9735\left[\frac{1}{2.7183^{\frac{1}{2}\left(\frac{X-37}{4}\right)^2}}\right]$
37	0.00	.0000	1.0000	1.0000	9.9735
38	.25	.0312	1.0317	.9692	9.6663
39	.50	.1250	1.1331	.8824	8.8015
40	.75	.2812	1.3247	.7548	7.5283
41	1.00	.5000	1.6487	.6065	6.0492
42	1.25	.7812	2.1842	.4578	4.5662
43	1.50	1.1250	3.0802	.4346	3.2379
44	1.75	1.5312	4.6239	.2162	2.1562
45	2.00	2.0000	7.3890	.1352	1.3494
46	2.25	2.5312	12.5692	.0795	.7928
47	2.50	3.1250	22.7598	.0439	.4382
48	2.75	3.7812	43.8708	.0227	.2273
49	3.00	4.5000	90.0171	.0111	.1107
50	3.25	5.2812	196.6454	.0050	.0507

$$y = \frac{Ni}{\sigma_x \sqrt{2\pi}} e^{-\frac{1}{2}\left(\frac{x - \mu_x}{\sigma_x}\right)^2}$$

$N = 100$
$i = 1$
$\mu_x = 37$
$\sigma_x = 4$

VALUE OF X

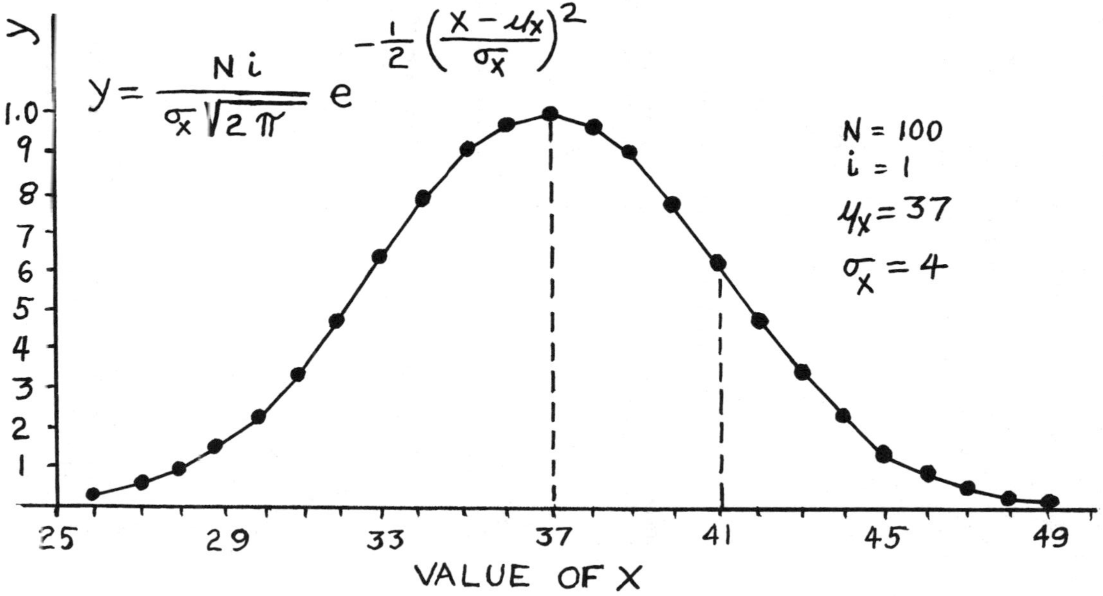

This is a graph of the values of Y for each X as calculated on the previous page.

As we see here, once you have specified the parameters in the equation for a normal curve you have specified a single unique function.

In this regard it is important to recognize that there are many functions that might bear some similarity to a given normal curve without in fact meeting all of the criteria for a normal function.

For example

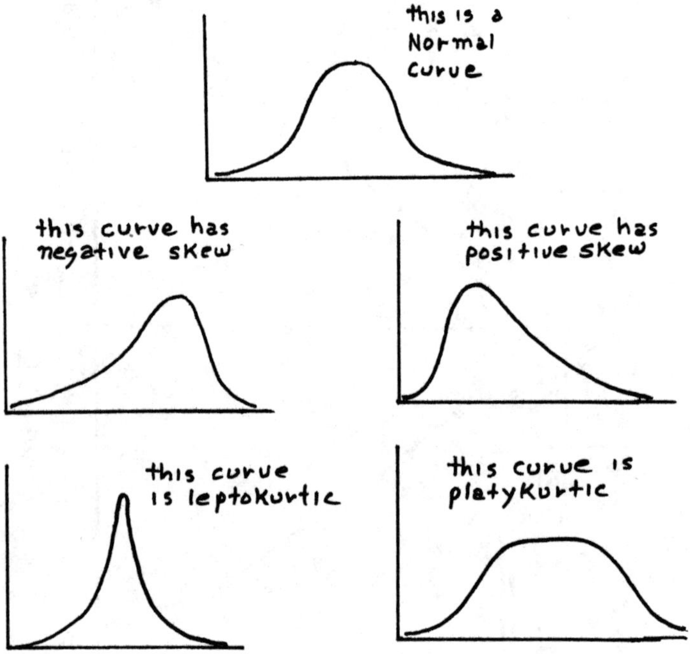
this is a Normal Curve
this curve has negative skew
this curve has positive skew
this curve is leptokurtic
this curve is platykurtic

Fortunately, however, as suggested earlier, many kinds of measures are in fact distributed normally and hence, we can use the equation for a normal curve to completely describe their distribution. In general, we tend to get normal distributions whenever we are dealing with measures that are affected by a very large number of chance factors. For example, the height of mature males in the U.S. tends to be distributed normally. Geneticists tell us that unlike eye color which tends to take on one of a few discrete hues and which is determined by a relatively small number of genes, a person's height is determined by a very large number of genes. Since the specific combination of genes carried by a given person depends upon the particular sets of genes donated by each parent and since the sets from each parent are determined randomly one would expect that across a large number of people, the physical expression of these combinations (height) would be normally distributed - as indeed it is. It's somewhat like tossing coins. Suppose that there are 1000 genes that go into determining a given person's height where each gene either contributes a finite increment to a person's height or it makes no contribution at all. This is like saying that a person's height is determined by tossing 1000 coins and counting the number of heads where each head contributes a finite increment to height. If we did so over and over for thousands of tosses and if we counted the number of heads on each toss we would get a distribution that was very similar to the following idealized histogram.

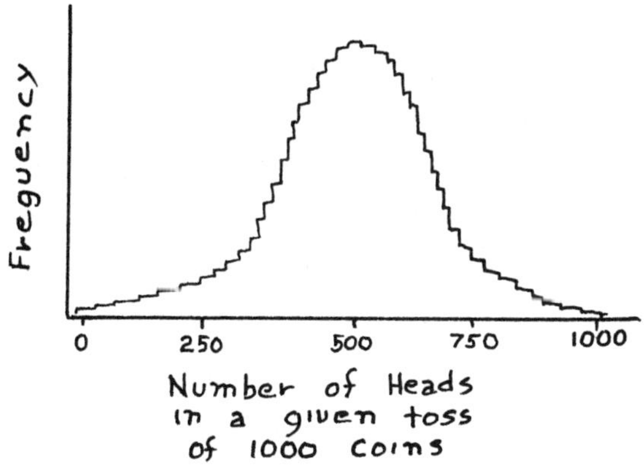

As can be seen, the most frequent combination is one containing 500 heads and accordingly the average height would be the one determined by 500 genes for height. Some people (tosses of 1000 coins) would have more than 500 genes for height and some would have less and over a large number of people the distribution would be approximately normal.

Errors in measurement tend to also be distributed normally. If you repeatedly measured the length of the desk before you to the nearest millionth of a centimeter the numbers you would record would tend to vary in some normal fashion with a given mean and variance. Once you have specified the distribution you could use the mean of the distribution as an index of the true length of the desk. Under such circumstances the variance of the distribution could be considered to be an index of the magnitude of your errors in measurement.

Because of the generality of the normal distribution, statisticians have gone to the trouble to determine the exact proportions of the normal distribution that fall on either side of an ordinate erected at various points along its abscissa. For example, since all normal distributions are symmetrical about the mean, an ordinate erected at the mean would divide any normal distribution into two equal areas.

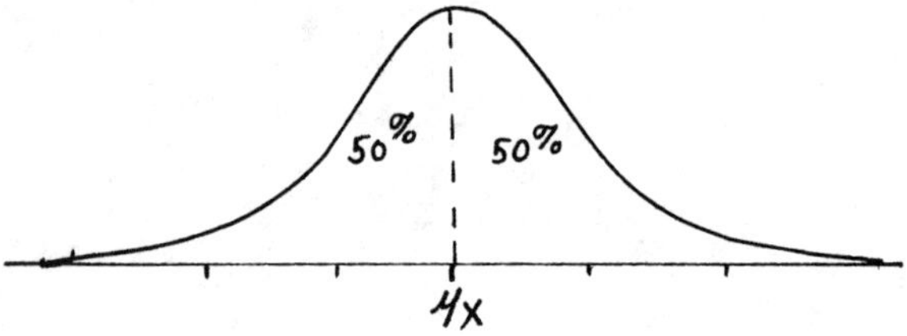

This, of course, would not be the case if the distributions were either positively or negatively skewed. But in such cases the distribution would not be normal. When our distribution is normal we know that if we erect an ordinate 1 σ_x above the mean it will divide the distribution into two areas, one of which consists of 84.13% of the total area.

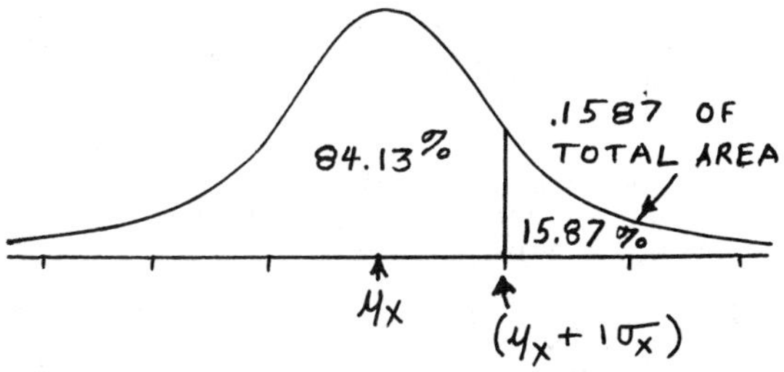

An ordinate erected 1.96 σ_X above the mean cuts off the upper 2.5% of the total area.

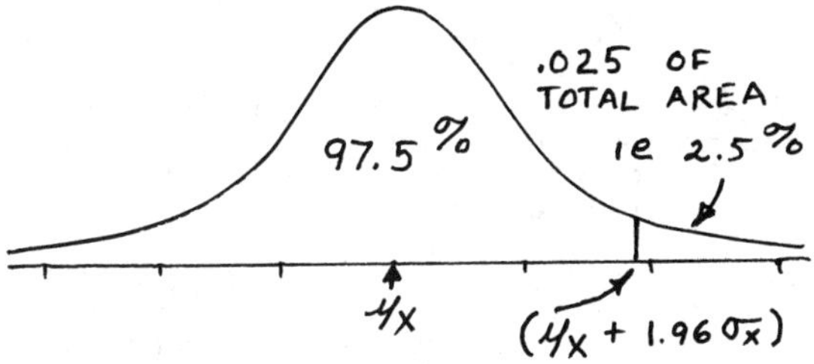

These statements are true for any normal distribution and if one ponders them for a moment it will become clear that they reflect the basic fact that when the equation for the normal curve specifies the precise height of the ordinate (Y) at each value of (X) it is also specifying the precise manner in which the area under the curve is distributed.

Thus, once the equation for a given normal curve is specified, statisticians can use integral calculus to specify the proportion of the total areas that fall under any given portion of that curve.

Of course it would be tremendously burdensome, indeed impossible, to specify the areas under various portions of all possible normal curves, because N, i, μ_X and σ_X can each have any of an infinite number of possible values. The way around this is to specify the areas under a normal curve with unit total area and with a mean = 0 and a variance equal to 1.

We saw earlier how any distribution can be transformed to a new distribtuion with mean = 0 and variance = 1 if we convert every item in the distribution into a Z score where

$$Z_X = \frac{X - \mu_X}{\sigma_X}$$

This means that one need only specify the areas under a normal curve with unit area and $\mu_X = 0$ and $\sigma_X = 1$ to to have a set of areas that would be characteristic of any normal curve.

The table of Z scores in the appendix of this book indicates the proportion of the total area in a normal curve that falls to the right of an ordinate erected at each of the indicated values of Z.

Suppose a student gets a score of 97 on an exam that was given nationwide. If the scores were normally distributed and if the mean of the distribution was 82 and its S.D. was 7, we would know (from the table of Z scores) that the student did equal to or better than 98.38% of the students that took the test. This follows because the raw score 97, when converted to a Z score, is

$$Z_{X=97} = \frac{97 - 82}{7} = 2.14$$

and from the table of Z scores we see that a Z score of 2.14 cuts off the upper .0162 of the distribution.

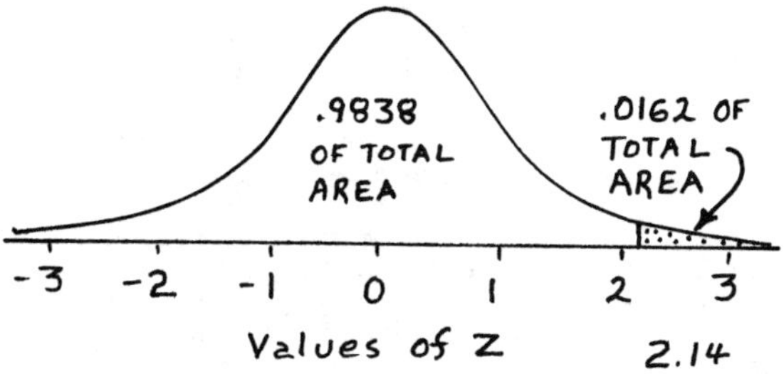

We have seen how one can use the normal curve to evaluate the probability of obtaining various values of X in drawing an item from a normal distribution. It seems important to again emphasize the point that if a given distribution is not normal we cannot use a normal function to evaluate this probability. Instead, we must use the function that in fact describes the distribution in question. Later we will have occasion to look at several distributions that are not normal, but for now we will largely be dealing with normal distributions.

Chapter 3

POPULATIONS AND SAMPLES

The Concept of a Population and the Concept of a Random Sample

Any set of items with a common observable characteristic can be regarded as a population (or universe). Once we have specified a given population we regard any set of items drawn from that population as a sample.

There are, of course, many ways that one could draw a sample from a given population For instance, we could arrange the items in the population in order of increasing magnitude and then select every other item. A systematic sample of this sort would, obviously be quite representative of the population. There are, however, many occasions when it is not possible to draw items in this kind of orderly fashion. Fortunately, on such occasions, one still has the opportunity to obtain a representative sample if each of the items in the sample is selected randomly.

Statisticians define a random sample as a set of items from a given population that are selected in such a way that each time an item is drawn, every item in the population has an equal opportunity to appear in the sample.

Consider a finite population that consists of the following seven numbers:

It might appear that we could draw a random sample of size 3 from this population by placing the items in a hat and blindly drawing out three items, one after the other.

Unfortunately, however, the three items obtained in this fashion would not in fact be a random sample because the opportunities for each item to appear in the sample were not in fact equal on each draw. Before choosing the first item the chance that a given item might be selected was 1/7, but once the first item was selected the probability that one of the remaining items would be selected had increased to 1/6 and by the time two items had been selected, each of the remaining items had a 1/5 chance of appearing in the sample.

We can avoid this difficulty by arranging to draw our items according to a slightly different procedure. As before, we start with a hat containing the seven items in the population. We then draw an item blindly and record its value. Then we replace the item, rescramble the set, and proceed to draw again. Once more, we record the value of the item, and after replacing it we go on to draw again. Clearly, with this procedure, on every draw each item in the population has an equal (1/7) opportunity to appear in the sample. It is of interest that with this procedure (called sampling with replacement), the size of the sample can, if we wish, exceed the size of the population. Thus, if we had reason to do so we could draw a sample of 1000 items from the finite seven item population we have just examined.

While these considerations make it clear that the only way one can actually draw a sample that is truly random is to use a replacement procedure, the criteria for a random sample will not be seriously violated if one draws items without replacement from a population that is extremely large relative to the size of the sample. This, of course, would invariably be the case when one's sample is drawn from a population that is infinite.

Finally, it is important to emphasize that while the criteria for a random sample requires that every item in the population has an equal opportunity to appear in the sample, these criteria do not require that every value in the population appear with equal frequency in the sample. Indeed, in the seven item population we have been examining, the odds of selecting an item with a value of 7 are three times as great as the odds of selecting an item with the value of say 5. It is this factor that explains why the distribution of values in a random sample is likely to be similar to the distribution of values in the parent population. Certainly, if we draw a large enough random sample we would be surprised if the relative frequencies of the values in the sample departed seriously from the relative frequencies of the values in the population. For example, in a large random sample from the above 7 item population we would expect to get approximately equal number of 5's, 6's, 8's and 9's, and we would expect to get approximately three times as many 7's. In other words we would be surprised if a histogram based on the data in the sample did not at least roughly approximate the histogram based on the data in the parent population.

The Sampling Distribution of $\overline{X}$ and the Central Limit Theorem

We have just seen why a random sample is likely to exhibit the characteristics of the population from which it was drawn. This factor enables us to use various measurements on a sample to estimate the comparable measurement on the parent population.

As noted earlier, statisticians refer to a given measurement on a population as a parameter of that population and they ordinarily use Greek letters to represent such measures. For example, we saw that the mean of a population is defined

$$\mu_X = \frac{\sum X}{N}$$

In the equation N = the number of items in the population

Statisticians use the term statistic to describe a given measure on a sample and they ordinarily use Latin letters to represent such measures. For example, the mean of a sample is specified by the statistic

$$\overline{X} = \frac{\sum X}{n}$$

In this equation n = the number of items in the sample

Because μ_X and $\overline{X}$ are both averages it might seem that one is being unnecessarily esoteric in using two different symbols to represent what are essentially identical sets of mathematical operations. It is important to do so, however, because it helps us to keep our thoughts in order when we get into the sometimes complex business of estimation.

If we are to understand how statistics are used to estimate parameters, we must understand the manner in which a given statistic will vary when one repeatedly draws random samples of a particular size (n) from a given population and calculates the statistic in question on each of the samples. Statisticians use the term sampling distribution to describe the distribution of values that is obtained in this fashion. Technically, a sampling distribution is defined as the set of values produced when one takes an infinite number of random samples of size n and calculates a given statistic on each of them. Obviously, we cannot actually carry out such a procedure in practice but there are a number of ways that we can visualize what would in fact happen if we did so.

To see how this is done we will begin by attempting to visualize the sampling distribution of the statistic

$$\bar{X} = \frac{\sum X}{n}$$

From our discussion thus far, it should be clear that the sampling distribution of X is the distribution of values that we would obtain if we took an infinite number of random samples of size n from a population with a given mean (μ_X) and a given variance (σ_X^2) and calculated X on each of the samples. The properties of the sampling distribution of X are specified by what statisticians call the central limit theorem. This important theorem consists of three statements which are presented here in summary form:

THE CENTRAL LIMIT THEOREM

(1) $\mu_{\bar{X}} = \mu_X$

(2) $\sigma_{\bar{X}}^2 = \sigma_X^2/n$

(3) If the original population is distributed normally, the sampling distribution of $\bar{X}$ will also be normal. Even if the original population is not normally distributed, the sampling distribution of $\bar{X}$ will approximate a normal distribution if the size of the samples (n) is large.

Let's examine each of the statements in the central limit theorem in detail:

Statement (1) ($\mu_{\bar{X}} = \mu_X$) asserts that the mean ($\mu_{\bar{X}}$) of the sampling distribution of $\bar{X}$ will equal the mean of the population (μ_X) from which the samples were drawn. We use the symbol $\mu_{\bar{X}}$ to describe the mean of the sampling distribution of $\bar{X}$ because the sampling distribution of $\bar{X}$ is conceived to contain an infinite number of $\bar{X}$s and, technically, such a distribution should be regarded as a population of $\bar{X}$s.

Statement (2) ($\sigma_{\bar{X}}^2 = \frac{\sigma_X^2}{n}$) asserts that the variance ($\sigma_{\bar{X}}^2$) of the sampling distribution of $\bar{X}$ will depend upon the variance of the population (σ_X^2) from which the samples were drawn and that it will also depend upon the number of items (n) in each of the samples. In this respect it is important to emphasize that a given sampling distribution of $\bar{X}$s always contains an infinite number of $\bar{X}$s, each of which is based on a sample containing exactly n items. In other words, the n in the equation $\sigma_{\bar{X}}^2 = \frac{\sigma_X^2}{n}$ refers to the size of the samples on which $\bar{X}$ is based. It does not refer to the number of samples (which is always infinitely large).

Statement (3) in the Central Limit Theorem asserts that when n is large, i.e., when we based our $\bar{X}$s on large enough samples) the sampling distribution of $\bar{X}$ can be described by the equation for a normal curve.

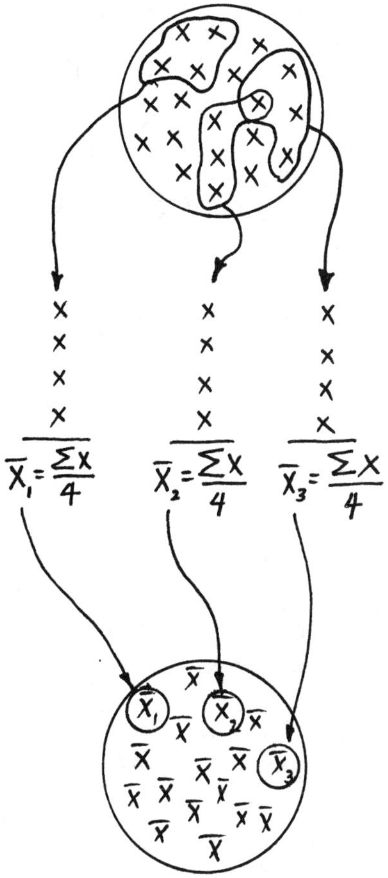

Here is a population of Xs.

Here are three of the infinitely large number of samples that make up a sampling distribution

Here is a sampling distribution of $\bar{X}$ based on samples of size n = 4

The central limit theorem tells us that even if we start with a population of that looks like this. →

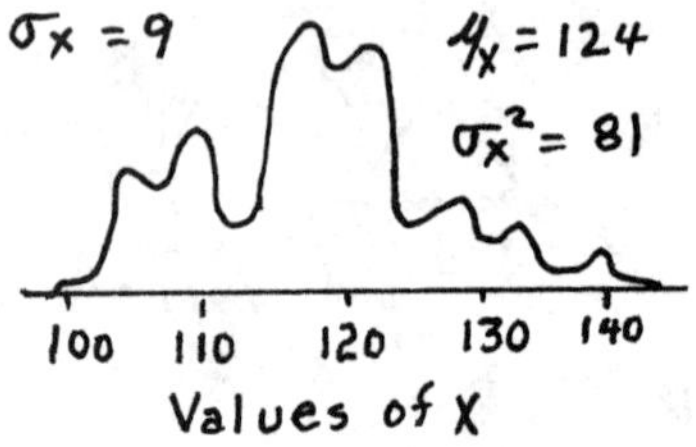

If we draw an infinite number of random samples of that are largen enough (for ex. n = 12) and if we calculate $\bar{X} = \frac{\Sigma X}{n}$ on each sample

The sampling distribution of $\bar{X}$ that we obtain will tend to look like this. →

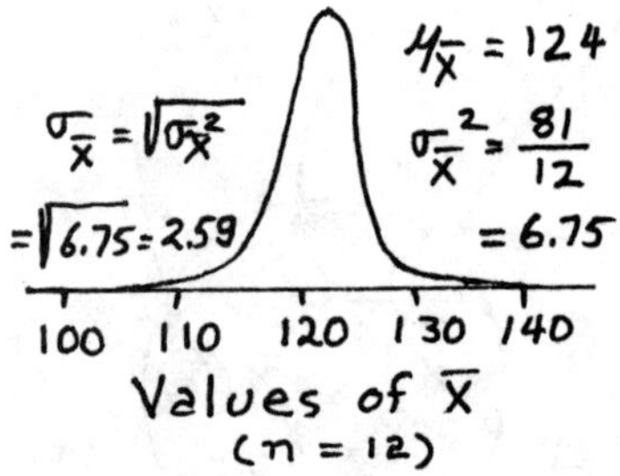

If we are to understand why the Central Limit Theorem is true, we will have to look very closely at what happens when we take random samples. To gain some insight into these effects let's examine what happens when we take random samples of size n = 2 from a population that consists of only 3 items. Of course we would be quite unlikely to ever have occasion to draw samples from such a curiously limited population, but our purpose here is to gain insight into the Central Limit Theorem and to do so it will be important to keep our numbers as simple as possible.

Here is a population of X_s. There are three of them: 3, 4, 5

$$\mu_X = \frac{1}{N} \sum X = \frac{1}{3}[3 + 4 + 5] = \frac{1}{3}[12] = \boxed{4}$$

$$\sigma_X^2 = \frac{1}{N} \sum (X - \mu_X)^2 = \frac{1}{3}[(3-4)^2 + (4-4)^2 + (5-4)^2]$$

$$= \frac{1}{3}[(-1)^2 + (0)^2 + (+1)^2]$$

$$= \frac{1}{3}[2] = \frac{2}{3} = \boxed{.66}$$

Let's imagine that we wish to form a sampling distribution of the statistic $(X_1 + X_2)$. To obtain a given value of $(X_1 + X_2)$ we draw a sample of two items (with replacement) and add together their values.

If we did this a large number of times we would discover that the following combinations of X_1 and X_2 would occur with equal frequency.

Value of X_1	Value of X_2	Value of $(X_1 + X_2)$
3	3	6
3	4	7
3	5	8
4	3	7
4	4	8
4	5	9
5	3	8
5	4	9
5	5	10

If we study this series of possibilities we see that there are several combinations of X_1 and X_2 that yield the same total. For example, when $X_1 = 3$ and $X_2 = 4$ we get the same value of $(X_1 + X_2)$ as when $X_1 = 4$ and $X_2 = 3$.

This means that if we took an infinite number of random samples of size n = 2 and calculated $(X_1 + X_2)$ on each we would obtain equal numbers of $(X_1 + X_2) = 6$ and $(X_1 + X_2) = 10$, we would obtain twice as many $(X_1 + X_2) = 7$, and we would obtain three times as many $(X_1 + X_2) = 8$.

The foregoing considerations make it clear that the sampling distribution of $(X_1 + X_2)$ would have the following form:

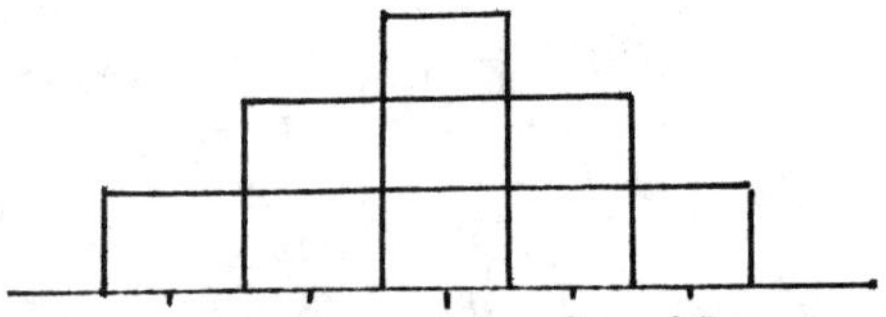

Of course a true sampling distribution of $(X_1 + X_2)$ would have an infinite number of $(X_1 + X_2)$s but since the relative frequencies of various values of $(X_1 + X_2)$ would be as shown above, the mean and variance of the above nine item distribution will be exactly the same as the mean and variance of the true sampling distribution of $(X_1 + X_2)$.

$$\mu_{(X_1+X_2)} = \frac{6+7+7+8+8+8+9+9+10}{9} = 8$$

$$\sigma^2_{(X_1+X_2)} = \frac{1}{9}\Big[(6-8)^2+(7-8)^2+(7-8)^2+(8-8)^2+(8-8)^2+(8-8)^2+(9-8)^2+(9-8)^2+(10-8)^2\Big]$$

$$= \frac{1}{9}[12] = \boxed{1.3\dot{3}}$$

Let's now compare the parameters of the sampling

distribution of $(X_1 - X_2)$ to the paramenters of the population of X_s from which the samples were drawn.

When we do so we discover that

$$\mu_{(X_1+X_2)} = \mu_{X_1} + \mu_{X_2}$$

$$= 4 + 4 = \boxed{8}$$

We also discover that

$$\sigma^2_{(X_1+X_2)} = \sigma^2_{X_1} + \sigma^2_{X_2}$$

$$= .6\dot{6} + .6\dot{6} = \boxed{1.3\dot{3}}$$

This points to an important insight about sampling. It implies that when we draw items randomly from a given population the mean of the sampling distribution of the sum of n items will be n times the mean of the population from which the items were drawn. Another way to say the same thing is to assert that

$$\mu_{(X_1+X_2+\cdots+X_n)} = \mu_{X_1} + \mu_{X_2} + \cdots + \mu_{X_n}$$

$$= n\mu_X$$

A similar phenomenon occurs with respect to the variance of the sampling distribution of sums of items. As we saw above

$$\sigma^2_{(X_1+X_2)} = \sigma^2_{X_1} + \sigma^2_{X_2}$$

If we extend our thinking to the more general case we can see that

$$\sigma^2_{(X_1+X_2+\cdots+X_n)} = \sigma^2_{X_1} + \sigma^2_{X_2} + \cdots + \sigma^2_{X_n}$$

$$= \boxed{n\sigma^2_X}$$

These two ideas can help us to understand why the first two statements in the Central Limit Theorem must be true.

To see how, consider that the mean of a given sample can always be expressed as follows:

$$\bar{X} = \frac{1}{n}(X_1 + X_2 + \cdots + X_n)$$

Accordingly, if we know the paramenters of the sampling distribution of ($X_1 + X_2 + \cdots + X_n$), we can use the rules for transforming Xs to determine the parameters of the sampling distribution of

$$\bar{X} = \frac{\Sigma X}{n} = \frac{1}{n}(X_1 + X_2 + \cdots + X_n)$$

From those rules it must be true that if

$$\mu_{(X_1 + X_2 + \cdots + X_n)} = n\mu_X$$

$$\mu_{\frac{1}{n}(X_1 + X_2 + \cdots + X_n)} = \frac{1}{n}(n\mu_X)$$

Just as

$$\mu_{cX} = c\mu_X$$

or in other words

$$\boxed{\mu_{\bar{X}} = \mu_X}$$

Moreover, if

$$\sigma^2_{(X_1 + X_2 + \cdots + X_n)} = n\sigma_X^2$$

$$\sigma^2_{\frac{1}{n}(X_1 + X_2 + \cdots + X_n)} = \frac{1}{n^2}(n\sigma_X^2)$$

Just as

$$\sigma^2_{cX} = c^2\sigma_X^2$$

or in other words

$$\boxed{\sigma^2_{\bar{X}} = \frac{\sigma_X^2}{n}}$$

We can gain further insight into these effects if we multiply each of the items in the above sampling distribution of ($X_1 + X_2$) by 1/2. When we do so we obtain the sampling distribution of $\bar{X}$ based on samples of size n = 2 that is generated when we taken our samples from a population consisting of the items 3, 4, and 5.

Here is that sampling distribution of $\bar{X}_3$.

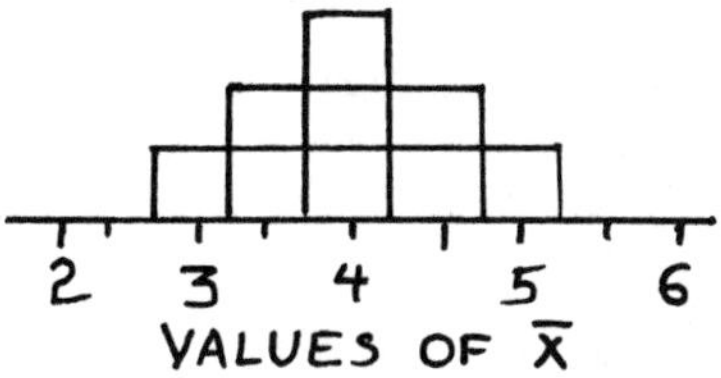

The mean and variance of this sampling distribution can be calculated directly as follows:

$$\mu_{\bar{X}} = \frac{1}{9}(3 + 3.5 + 3.5 + 4 + 4 + 4 + 4.5 + 4.5 + 5) = \frac{1}{9}(36)$$

$$\boxed{\mu_{\bar{X}} = 4}$$

$$\sigma^2_{\bar{X}} = \frac{1}{9}\Big[(3-4)^2 + (3.5-4)^2 + (3.5-4)^2 + (4-4)^2 + (4-4)^2 + (4-4)^2 + (4.5-4)^2 + (4.5-4)^2 + (5-4)^2\Big]$$

$$= \frac{1}{9}(3)$$

$$\boxed{\sigma^2_{\bar{X}} = .3\dot{3}}$$

Clearly, these values are identical to the values

specified by the Central Limit Theorem. According to that theorem when we take samples of size n = 2 from a population with

$$\mu_x = 4$$

and $$\sigma_x^2 = .6\dot{6}$$

the mean of the sampling distribution of $\bar{X}$ will equal μ_x

i.e. $$\mu_{\bar{X}} = 4$$

and the variance of the sampling distribution of $\bar{X}$ will equal σ_x^2/n

i.e., $$\sigma_{\bar{X}}^2 = \frac{.6\dot{6}}{2} = .3\dot{3}$$

In addition to exemplifying the rationale for the first two assertions in the Central Limit Theorem, this numerical example provides an indication of why, even when sampling from a population that is not normal, the sampling distribution of means tends to increasingly approximate a normal distribution as the sample size (n) increases.

Consider the distribution that describes the three items in the limited population we have been studying. Properly speaking, this is a rectangular distribution because the values 3, 4, and 5 occur with equal frequency.

Now, consider the sampling distribution of $\bar{X}$ based on n = 2 that we just derived. Even with an n as small as 2, the sampling distribution of $\bar{X}$ is much more like a normal curve than the rectangular distribution that describes the original population.

In summary then, the example we have been studying very nicely exemplifies all three facets of the Central Limit Theorem.

1) $\mu_{\bar{X}} = \mu_X$

2) $\sigma_{\bar{X}}^2 = \dfrac{\sigma_X^2}{n}$

3) The sampling distribution of $\bar{X}$ tends to increasingly approximate a normal distribution as n increases.

Transforming a Sampling Distribution of $\bar{X}$s into z Scores

Consider a normal distribution of Xs with mean = μ_X and variance = σ_X^2

From the Central Limit Theorem we know that the sampling distribution of $\bar{X}$ based on samples of size n will have the following form:

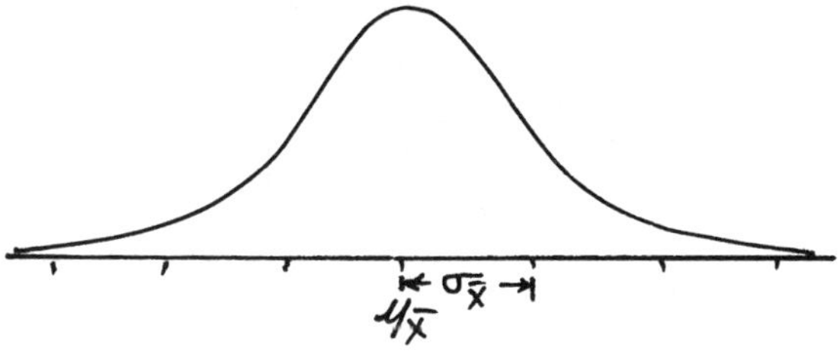

$$\mu_{\bar{X}} = \mu_X \qquad \sigma_{\bar{X}} = \sqrt{\sigma_X^2/n} = \frac{\sigma_X}{\sqrt{n}}$$

Earlier we learned that we can transform any distribution of Xs into a new distribution with mean = 0 and standard deviation = 1 if we first subtract μ_X from every item and then divide each of the resulting ($X - \mu_X$)s by σ_X. That is, we can transform any distribution of Xs into a new distribution with mean = 0 and standard deviation = 1 if we transform every X into a z_X , where

$$z_X = \frac{X - \mu_X}{\sigma_X}$$

We can also transform any sampling distribution of $\bar{X}$s into a new distribution with mean = 0 and standard deviation = 1 if we transform each $\bar{X}$ into a $z_{\bar{X}}$,where

$$z_{\bar{X}} = \frac{\bar{X} - \mu_{\bar{X}}}{\sigma_{\bar{X}}}$$

Here is another (equivalent) formula for $z_{\bar{X}}$ that also gives us a distribution of $z_{\bar{X}}$ with

$$\mu_{z_{\bar{X}}} = 0 \quad \text{and} \quad \sigma_{z_{\bar{X}}} = 1$$

$$z_{\bar{X}} = \frac{\bar{X} - \mu_X}{\sqrt{\frac{\sigma_X^2}{n}}}$$

To obtain this formula we merely made use of the equivalences expressed by the Central Limit Theorem

namely: $\mu_{\bar{X}} = \mu_X$

and $\sigma_{\bar{X}}^2 = \frac{\sigma_X^2}{n} \quad \therefore \quad \sigma_{\bar{X}} = \sqrt{\frac{\sigma_X^2}{n}}$

The formula

$$Z_{\bar{X}} = \frac{\bar{X} - \mu_X}{\sqrt{\frac{\sigma_X^2}{n}}}$$

is important because it provides us with a convenient way of assessing the probability of obtaining various values of $\bar{X}$ when our sample comes from a normal population with a given mean and a given variance.

For example, given a normal population with μ_X = 350 and σ_X^2 = 160 we can use the statistic

$$Z_{\bar{X}} = \frac{\bar{X} - \mu_X}{\sqrt{\frac{\sigma_X^2}{n}}}$$

to determine the probability of drawing a random sample of size n = 10 with mean ($\bar{X}$) equal to or greater than 358. To do so we convert the $\bar{X}$ in question, i.e., $\bar{X} = 358$ into a $Z_{\bar{X}}$ as follows:

$$Z_{\bar{X}} = \frac{358 - 350}{\sqrt{\frac{160}{10}}} = \frac{8}{4} = \boxed{+2}$$

We then look up the area in the appended Z score table to determine the region of the normal curve that is bounded by the obtained Z score.

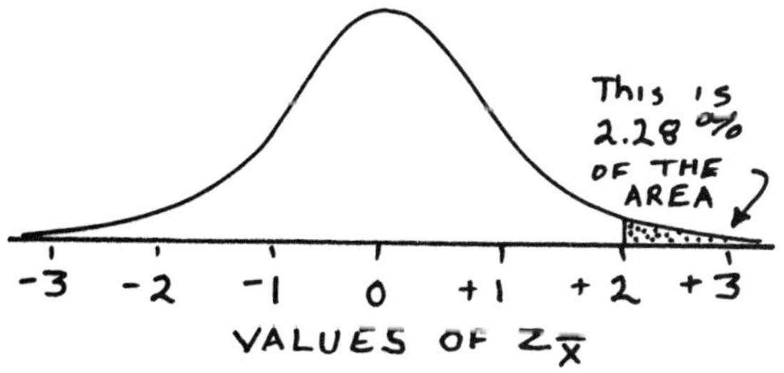

When we do so we discover that the area in the upper tail bounded by $Z_{\bar{X}} = +2$ is 2.28% of the total distribution. Accordingly, we can conclude that the probability of obtaining a sample of size n = 10 with a mean ($\bar{X}$) equal to or greater than 358 from the population in question is .0228. In this regard, it is important to recognize that our conclusion is based on our knowledge that the original population of Xs was normally distributed. Clearly, if the original population was not normally distributed we would not be justified in using the table of Z scores to ascertain probability unless we had some reason to believe that our sample size was large enough to insure that the sampling distribution of $Z_{\bar{X}}$ would nonetheless approximate a normal distribution. As might be expected, the sample size required for this to occur will depend upon the nature and size of the departure from normality in the original population.

Chapter 4

TESTS OF THE HYPOTHESES ABOUT μ_X WHEN σ_X^2 IS KNOWN

Once we have mastered the concept of sampling disribution and we understand the important message carried by the central limit theorem and once we know how to transform scores we are in a position to examine various procedures for testing statistical hypotheses.

Briefly stated, a statistical hypothesis is an assertion about a parameter of a particular population.

The test of a statistical hypothesis is a procedure for deciding, with a specified degree of confidence, whether or not to reject the assertion. In testing a statistical hypothesis we base our decision on the information derived from a random sample that is drawn from the population with which the hypothesis is concerned. Consider, for example, the following problem.

We have before us a population of Xs. We know that the distribution is normal and that its variance is $\sigma_X^2 = 9$. We do not, however, know the value of the mean of the population, i.e., $\mu_X$is unknown. Suppose, further, that we have drawn a random sample of 10 items from the population and we wish to use the information in this sample to test the statistical hypothesis that $\mu_X = 35$.

In short, we wish to test the assertion that the mean of the population from which the sample was drawn is equal to 35. Finally let's suppose that we desire to test this hypothesis, under circumstances where the probability of rejecting the hypothesis, if it is true is only .05.

To see how to carry out the test let's assume that the hypothesis $\mu_X = 35$ is in fact true. We have just seen (in the previous chapter) that when we draw samples from a normal population with the mean = μ_x and variance = σ_x^2 the sampling disrtibution of the statistic

$$Z_{\bar{X}} = \frac{\bar{X} - \mu_x}{\sqrt{\frac{\sigma_x^2}{n}}}$$

is normal with mean = 0 and variance = 1. This means that if the hypothesis is true the sampling distribution of the statistic

$$Z_{\bar{X}} = \frac{\bar{X} - 35}{\sqrt{\frac{9}{10}}}$$

will be normal with $\mu_{Z\bar{X}}$ = 0 and $\sigma_{Z\bar{X}}$ = 1. From the appended table of Z scores we discover that the values of Z that cut off the upper and lower 2.5% of such a distribution are Z = +1.96 and Z = -1.96

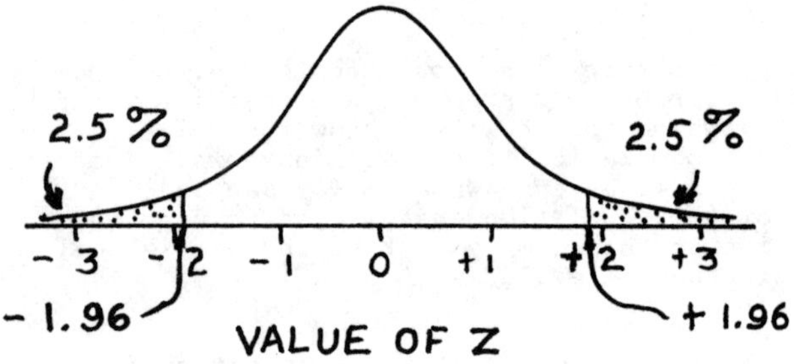

This tells us that if the hypothesis we are testing is true, the chances of getting a sample that will yield a $Z_{\bar{X}}$ that is either > +1.96 or < -1.96 is exactly .05 i.e. 2.5% x 2 = 5%.

Suppose the mean of the one sample before us is $\bar{X}$ = 34

In this case $Z_{\bar{X}} = \dfrac{34 - 35}{\sqrt{\dfrac{9}{10}}} = \dfrac{-1}{.94868}$

$= -1.054$

which is $-1.96 < -1.054 < +1.96$

Clearly if our sample yields a $Z_{\bar{X}}$ with this value we would have no basis to reject the hypothesis tested. This follows because as seen below samples of size 10 that yield values of $-1.96 < Z_{\bar{X}} < +1.96$ would be expected 5% of the time when the hypothesis tested is true.

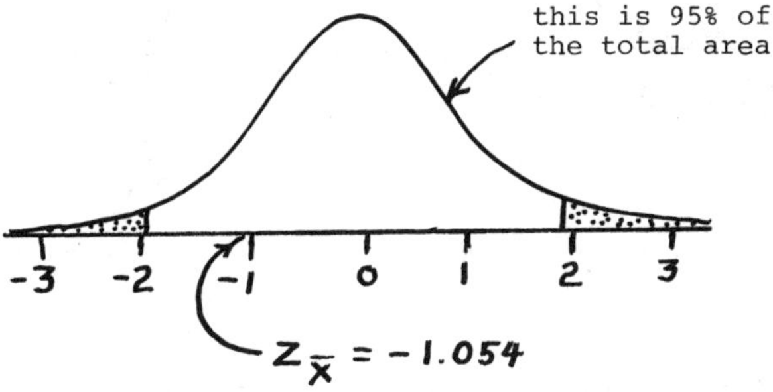

$Z_{\bar{X}} = -1.054$ is among those values that occur 95% of the time when the hypothesis tested is true.

Suppose, however, that the sample mean had been = 38

In this case

$$Z_{\bar{X}} = \frac{38 - 35}{\sqrt{\dfrac{9}{10}}} = +2.8$$

and we would be justified in rejecting the hypothesis

that $\mu_{\bar{x}} = 35$.

This would follow because as seen below a value of $z_{\bar{x}} = 2.8$ is among those extreme values of $z_{\bar{x}}$ that would only occur 5% of the time when the hypothesis tested is true.

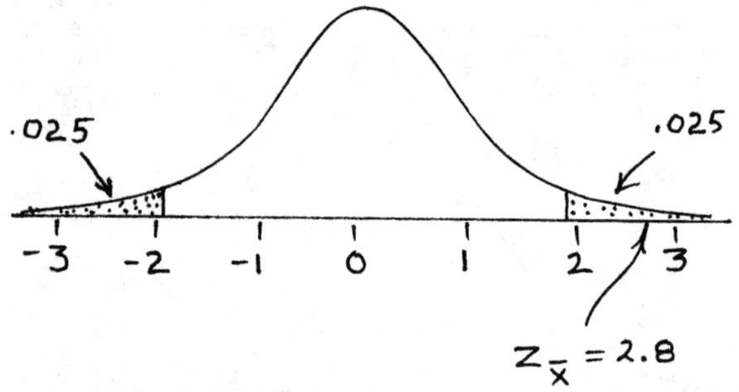

Let's review and summarize these ideas.

We have defined a statistical hypothesis as a statement about the value of a parameter of a population.

The test of a statistical hypothesis is a method for deciding, with a specified degree of confidence, whether or not to reject the hypothesis.

In testing a statistical hypothesis we based our decision upon the information derived from a random sample from the population with which the hypothesis is concerned.

To test a given statistical hypothesis we carry out the following sequence of operations.

(1) State the hypothesis.

(2) Decide upon the risks we are willing to

take. There are two such risks.

I. We might reject the hypothesis when it is in fact true.

Note: The probability of rejecting the hypothesis when it is true is called α (alpha). We control this

probability by setting α at any value we choose. Naturally, the smaller we make α the less likely you are to reject a true hypothesis. It is, perhaps, not so obvious, however, that the smaller we make α the more stringent our test and the more likely we are to accept the hypothesis when it is in fact false. This leads us to the second kind of risk.

II. We might fail to reject the hypothesis when it is false, i.e., when some alternative hypothesis is true.

(3) Choose a statistic to employ in testing the hypothesis.

Note: A statistic is a measure on a sample. Generally, the statistic used will depend upon the hypothesis we are testing and the information we have about the population from which the sample was drawn. $z_{\bar{x}}$ is an example of a statistic that can be used to test hypotheses about μ_x when the parent population is normal and σ_x^2 is already known.

(4) Determine the sampling distribution that the statistic would have if the hypothesis were true and locate the rejection region (or regions).

Note: Although the location of the rejection region will depend upon a number of considerations, the total area of this region must equal the value that we have chosen for α .

(5) Compute the statistic in question on the sample at hand and determine whether or not it falls in the rejection region.

Note: In computing the sample statistic we do so under the assumption that it will represent a sample of one item from the sampling distribution that we would obtain if the hypothesis were true and we took repeated samples and computed the appropriate statistic each time.

There are only two reasons why the obtained statistic might fall in the rejection region.

I. The hypothesis is true and the sample statistic represents a random occurrence with a probability equal to α .

II. The hypothesis is false.

If the sample statistic does fall in the rejection region, we say that the reason was II rather than I; in other words we reject the hypothesis.

Note that we have conducted our test in such a way that the probability of rejecting the hypothesis, if it were true, was precisely equal to alpha. We say, therefore, that we reject at the α level of confidence.

If the sample statistic does not fall in the rejection region, we have no basis upon which to reject the hypothesis.

By the procedures outlined here we have no notion of the risks involved in failing to reject the hypothesis. We do, however, know (and control) the risks in rejecting it. It is for this reason that research workers often prefer to state their statistical hypothesis in such a way that its rejection can be taken as evidence in support of the findings. This is not to say that there are no ways of specifying the risks involved in failing to reject a given hypothesis, but as will be elaborated later in this book, to do so we must be able to state an alternative hypothesis and dealing with the complexities that this introduces seems best postponed until the procedures outlined here are more thoroughly understood.

A practical example may help to clarify the strategy employed here. Suppose that a manufacturer of light bulbs wishes to determine if his production equipment is functioning properly. The equipment is designed to produce bulbs with a burning life of 740 hours. Of course even the best of equipment is imperfect and his is no exception. Thus, even when the equipment is working perfectly it produces some bulbs that burn out in less than 740 hours and some that take more than 740 hours to expire. In fact, when functioning exactly according to specification the burning times of the bulbs produced by the equipment is normally distributed with mean = 740 hours and $\sigma_{\bar{X}}$ = 20 hours

Let's suppose further that the manufacturer knows from experience that when the equipment malfunctions it tends to produce bulbs with either increased or decreased average burning times, but the variability among bulbs is not ordinarily affected.

Since we know that the burning times are distributed normally, we can use a $z_{\bar{X}}$ test to help the manufacturer decide whether or not the equipment is working according to specification. Our strategy would be to take a random sample of bulbs from the production line, and then carry out a test of the statistical hypothesis that the sample came from a population with a mean of 740 hours.

Suppose that for practical purposes we decide that our sample size will be n = 25. Let's suppose further that the manufacturer has agreed that he is willing to take a .05 risk of concluding that the equipment is malfunctioning when in fact it is functioning properly (i.e., he agrees to set α = .05) and that he is just as concerned with detecting malfunctions in both directions. That is, he knows that if the average burning times of the product is too low, he will have dissatisfied customers and if the average burning time is too high he'll ultimately lose sales. Of course, we may disagree with this implication of planned obsolescence, but it's nonetheless a factor we must consider in our statistical test.

Once we have come to grips with these issues we are prepared to carry out the test. To do so, we calculate the average burning time of the 20 bulbs in our sample. Let's suppose that when we do so we obtain

$$\bar{X} = \frac{\sum X}{n} = \frac{18200}{25} = 728$$

Next we enter this value into the formula for

$$z_{\bar{X}} = \frac{\bar{X} - \mu_h}{\sqrt{\frac{\sigma_X^2}{n}}} = \frac{728 - 740}{\sqrt{\frac{400}{25}}} = \frac{-12}{4} = -3$$

From the table of Z scores we find that the Z scores that cut off the upper and lower .025 of the distribution are Z = +1.96 and Z = -1.96.

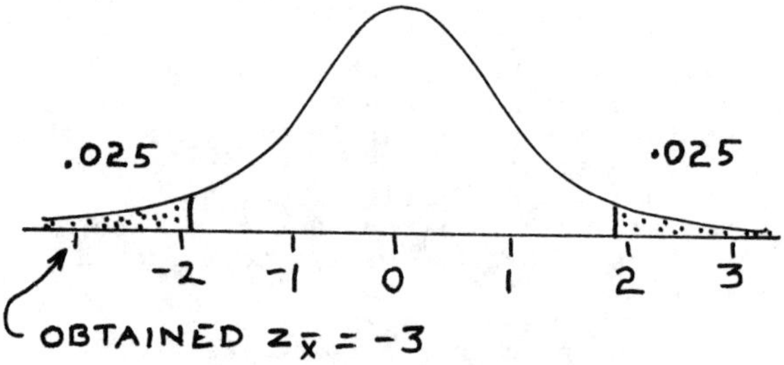

This is the sampling distribution of $z_{\bar{X}}$ that would be obtained if the hypothesis tested ($\mu_h = 740$) were true.

Since the obtained $z_{\bar{X}}$ is -3, a value which is among those extreme values of $z_{\bar{X}}$ that would only occur 5% of the time when the statistical hypothesis is true, we conclude that the equipment is producing substandard bulbs.

This example is summarized on the next page:

Example I - Testing an exact hypothesis about the mean, σ_X^2 known.

Illustration

In order to determine if certain equipment is producing substandard light bulbs we test the hypothesis that the equipment is producing light bulbs of standard quality. (Average burning time equal to some specified value (μ_h). Variance of burning time is assumed known.

Method

(1) Hypothesis: Population mean equal to some specified value (μ_h). (Given: population variance = σ_X^2)

(2) Set α

(3) Use

$$Z_{\bar{X}} = \frac{\bar{X} - \mu_h}{\sqrt{\frac{\sigma_X^2}{n}}}$$

where

n = Sample size

$\bar{X}$ = Sample mean

μ_h = Hypothesized population mean

σ_X^2 = Known population variance

(4)

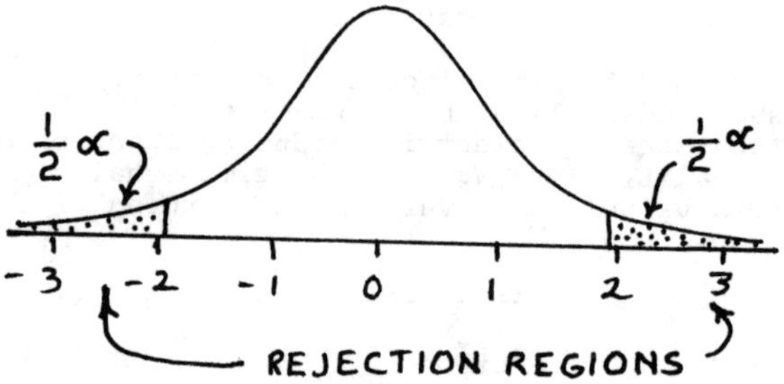

We determine the sampling distribution of $Z_{\bar{X}}$ under the assumption that the hypothesis is true and we specify the location of the rejection region (or regions).

(5) Compute $Z_{\bar{X}}$ and determine whether or not it falls in the rejection region.

Notice that in the illustration if we can reject the hypothesis tested we will have reason to believe that the output of the equipment is substandard.

Chapter 5

THE RATIONALE FOR USING $s_x^2 = \frac{\sum(X-\bar{X})^2}{n-1}$ TO ESTIMATE σ_x^2.

We have just seen how the logic of statistical reasoning leads us to a procedure for testing the hypothesis that the mean of a given population (μ_x) is some specified value. To carry out that test, however, we had to know the value of σ_x^2 (the variance of the population in question). In most practical situations this information is unavailable. Fortunately, however, under such circumstances we can use the information in our sample to get an estimate of the variance of the population. The symbol most statisticians use to represent an estimate of a population variance based on the information in a sample is s_x^2.

We just saw how we can employ the statistic

$$z_{\bar{X}} = \frac{\bar{X} - \mu_h}{\sqrt{\sigma_x^2/n}}$$

to test a hypothesis about μ_x when σ_x^2 is known. We use a comparable statistic

$$t = \frac{\bar{X} - \mu_h}{\sqrt{s_x^2/n}}$$

to test a hypothesis about μ_x when we have to estimate σ_x^2 from the information in our sample. In general, our testing procedure follows the same format. We state our hypothesis and decide upon a statistic that will be sensitive to departure from that hypothesis. Once we have settled upon the appropriate statistic to use, we determine the sampling distribution that the statistic would have if the hypothesis were true. If our sample yields a value of the statistic (in this case t) that is among the extreme values that occur infrequently when the hypothesis is true, we reject the hypothesis. Otherwise we have no basis to reject it.

The term s_x^2 in the formula for t is specified algebraically as follows:

$$s_x^2 = \frac{\sum(x-\bar{x})^2}{n-1}$$

To calculate s_x^2 on a given sample we first calculate the mean of the sample $\bar{X} = \frac{\sum X}{n}$. Then we subtract $\bar{X}$ from each item in the sample and square the resulting quantity. We next add together the several values of $(X - \bar{X})^2$ and divide this sum by one less than the size of the sample (i.e., we divide by n - 1).

As can be seen s_x^2 is not the variance (σ_x^2) of the set of items that make up the sample. The variance of a given set of items is a measure on the set that is used to describe their dispersion. In short, the variance (i.e., the average of the squared deviations from the mean) is a parameter that tells us about one of the characteristics of the specific set of items before us. s_x^2 on the other hand, is a measure on a set of items that is used to estimate the σ_x^2 of the parent population from which the set (sample) was drawn (i.e., like $\bar{X}$, s_x^2 is a statistic).

The rationale for using the statistic s_x^2 to estimate the parameter (σ_x^2) of the parent population is quite complicated and its development will require the introduction of several new concepts. In order to keep the logical steps as small and as straightforward as possible, each new concept will be explained as the need for it arises. An inevitable consequence of this practice will be to make the account rather lengthy and this means that its basic structure will be somewhat elusive. But, as will be seen, all of the concepts hang together and if the student carefully thinks through each of them the logical structure of the material will become increasingly clear and, what is equally important, the student will acquire a good grasp of its several basic concepts. We will find that as we proceed to examine the logic of statistical reasoning we will have many occasions to make use of these concepts. So, if the logic of what we are about to examine seems difficult to grasp at first, persevere in your efforts to master it. After all, mathematics has been around for thousands of years, whereas the logic of statistical reasoning has only begun to become apparent in the past few decades. If this logic has eluded mankind for thousands of years, it would seem presumptuous to assume that it would be easy to grasp.

Suppose we begin with a finite population of Xs with mean = μ_X and variance = σ_X^2. That is we have before us a set of values:

$$X_1, X_2, X_3 \cdots X_N$$

with $$\mu_X = \frac{\Sigma X}{N}$$

and $$\sigma_X^2 = \frac{\Sigma(X - \mu_X)^2}{N}$$

Here is that population of Xs.

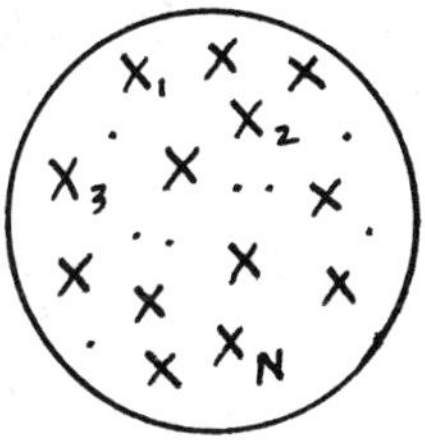

Suppose now that we subtract μ_X from every item in this population and then square each of the resulting quantities. This will yield a new set (population) of N values:

$$(X_1 - \mu_X)^2, (X_2 - \mu_X)^2 \cdots (X_N - \mu_X)^2$$

Here is the population of $(X-\mu_X)^2$s

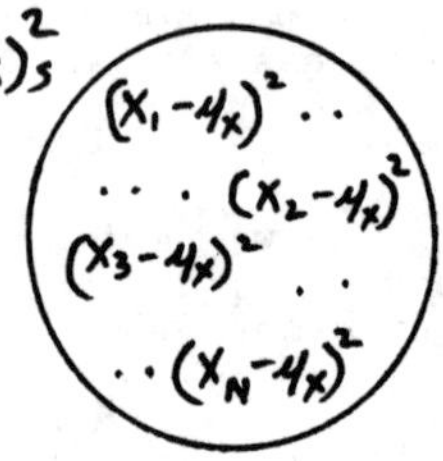

Like any other set of values, this set must have a mean which is equal to the sum of the set divided by the number of items in the set.

Algebraically, the mean of this set is described as follows:

$$\mu_{(X-\mu_X)^2} = \frac{\sum (X-\mu_X)^2}{N}$$

Now this is a very interesting turn of events because we see that the parameter that describes the mean of this population ($\mu_{(X-\mu_X)^2}$) is exactly equal to the parameter (σ_X^2) that describes the variance of the initial set of Xs.

i.e. $$\mu_{(X-\mu_X)^2} = \frac{\sum (X-\mu_X)^2}{N} = \sigma_X^2$$

Suppose now that we take a random sample of n items from this population

Note that there are n items in the sample

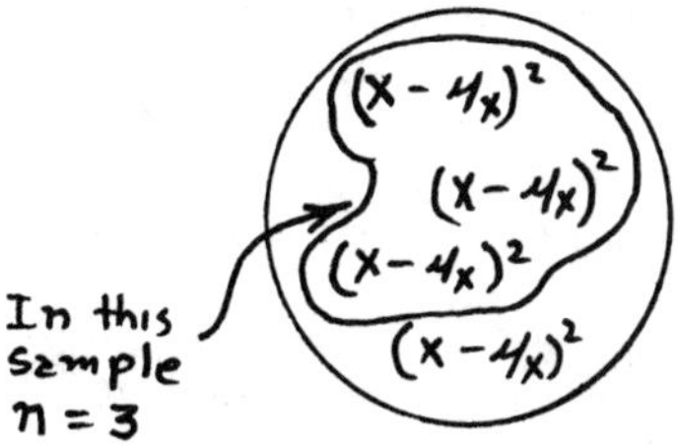

Note that there are N items in the population

The mean of this sample is $\dfrac{\sum (X - \mu_X)^2}{n}$

We know (from the Central Limit Theorem) that the mean of a given sample can be used to estimate the mean of the population from which the sample was drawn. This is because the Central Limit Theorem tells us that the mean of the sampling distribution of means is equal to the mean of the parent population.

When applied to the present situation this principle tells us that the mean of a given sample of $(X - \mu_X)^2$'s, i.e., the statistic

$$\frac{\sum (X - \mu_X)^2}{n}$$

computed on that sample, can be used to estimate the mean of the population of $(X - \mu_X)^2$'s from which the sample was drawn.

We saw earlier that

$$\mu_{(X - \mu_X)^2} = \sigma_X^2$$

It must follow therefore, that the mean of a sample of $(X - \mu_X)^2$'s from a population of $(X - \mu_X)^2$s can be used as an estimate of the σ_X^2 of the population of Xs from which the $(X - \mu_X)^2$s were derived.

A statistician would use the expression

$$\frac{\sum (X - \mu_X)^2}{n} \longrightarrow \sigma_X^2$$

to summarize this fact. In ths expression the arrow ($\longrightarrow$) is interpreted to imply that the mean of the sampling distribution of the statistic on the left

$\frac{\sum (X - \mu_X)^2}{n}$ is exactly equal to the quantity on the

right (σ_X^2). Another way to say the same thing is to assert that the statistic on the left of the arrow is an unbiased estimate of the parameter on the right. By this terminology the mean of the sampling distribution of a biased statistic would not equal the parameter that the statistic is used to estimate.

For example, the mean of the sampling distribution of the statistic $2\frac{\sum X}{n}$ does not equal μ_X (it equals $2\mu_X$). A statistician would express this fact by asserting that $2\frac{\sum X}{n}$ is a biased estimate of μ_X. Another way that a statistician might make the same statement would be to use the shorthand expression:

$$2\frac{\sum X}{n} \not\longrightarrow \mu_X$$

Let's now again consider the expression:

$$\frac{\sum(X - \mu_X)^2}{n} \longrightarrow \sigma_X^2$$

This is an extremely important statement, because it tells us that if we have a population of Xs with a given mean (μ_X) and a given variance (σ_X^2), we can obtain an unbiased estimate of that variance (σ_X^2) if we draw a random sample of n items and calculate the statistic

$$\frac{\sum(X - \mu_X)^2}{n}$$

on that sample.

To carry out this calculation we would merely subtract μ_X from each X in the sample, square the resulting $(X - \mu_X)$ and then divide the sum of the $(X - \mu_X)^2$s by n (the size of our sample).

Of course, showing that the statistic

$$\frac{\sum(X - \mu_X)^2}{n} \longrightarrow \sigma_X^2$$

is not itself a proof of the proposition that we set out to demonstrate: namely that the statistic

$$s_X^2 = \frac{\sum(X - \bar{X})^2}{n-1} \longrightarrow \sigma_X^2$$

but it's a step in the right direction. Unfortunately, however, before continuing in that direction we have to sidetrack a bit and look at another kind of sampling distribution.

Again, suppose we have a population of Xs with mean = μ_X and variance = σ_X^2 .

Here is that population

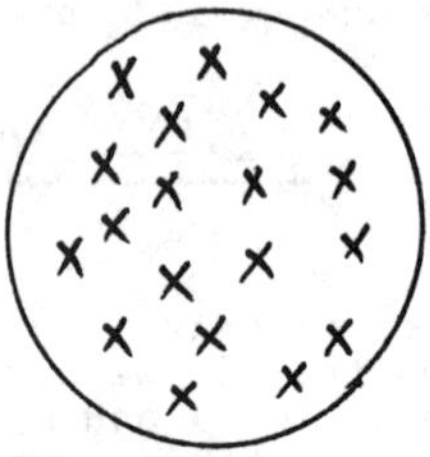

mean = μ_X
variance = σ_X^2

Now imagine that we repeatedly draw random samples of size n and compute $\frac{\sum X}{n}$ on each sample, i.e., we form a sampling distribution of $\bar{X}$.

Here is that sampling distribution

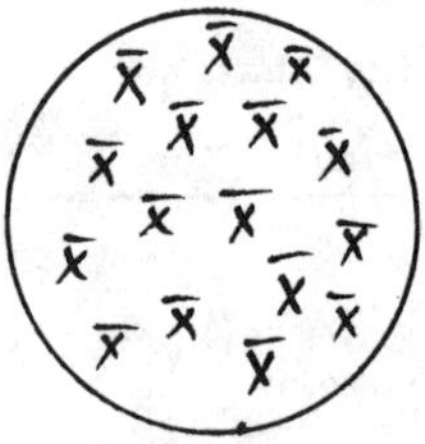

From the Central Limit theorem we know that

$$\mu_{\bar{X}} = \mu_X \quad \text{and} \quad \sigma_{\bar{X}}^2 = \frac{\sigma_X^2}{n}$$

Now imagine that we subtract   from each of the items (   ) in this sampling distribution and then square each of the resulting quantities.

Here is the set of $(\bar{X} - \mu_{\bar{X}})^2$s that would result

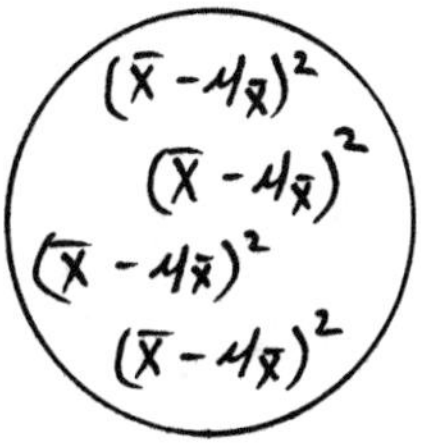

Earlier we saw that the mean of a population of $(X - \mu_X)^2$s was equal to the σ_X^2 of the original set of Xs from which the population was derived. If we apply a similar logic to the present situation it should become apparent that the mean of the set of $(\bar{X} - \mu_{\bar{X}})^2$s depicted above must be the variance ($\sigma_{\bar{X}}^2$) of the sampling distribution of means ($\bar{X}$ s) from which the population of $(\bar{X} - \mu_{\bar{X}})^2$s was derived. It should also become apparent that if we randomly select one of the $(\bar{X} - \mu_{\bar{X}})^2$s from this population its value would be an unbiased estimate of $\sigma_{\bar{X}}^2$

This follows from what we know about sampling. After all, the set of $(\bar{X} - \mu_{\bar{X}})^2$s is merely a set of numbers and we know that the mean of the sampling distribution of means of samples of a given size is always equal to the mean of the population from which the samples were drawn. The only novel feature in the present situation is that our sample size is n = 1, i.e., we have only one $(\bar{X} - \mu_{\bar{X}})^2$, but there is nothing about the Central Limit Theorem that says that n cannot = 1.

All of this leads to the conclusion that

$$(\bar{X} - \mu_{\bar{X}})^2 \rightarrow \sigma_{\bar{X}}^2$$

But if $(\bar{X} - \mu_{\bar{X}})^2 \rightarrow \sigma_{\bar{X}}^2$ it must also be true that

$$(\bar{X} - \mu_X)^2 \rightarrow \frac{\sigma_X^2}{n}$$

This is because we know (from the central limit theorem) that $\mu_{\bar{X}} = \mu_X$ and $\sigma_{\bar{X}}^2 = \frac{\sigma_X^2}{n}$

In other words, we have just shown that if we draw a random sample of size n from a population of Xs with mean = μ_X and variance = σ_X^2.

The statistic $(\bar{X} - \mu_X)^2$ calculated on that sample is an unbiased estimate of $\frac{\sigma_X^2}{n}$. This fact in combination with the material we covered earlier gives us two important assertions. 1) For a given sample of Xs the statistic

$$\frac{\sum (X - \mu_X)^2}{n} \rightarrow \sigma_X^2$$

and 2) for the same sample of Xs the statistic

$$(\bar{X} - \mu_X)^2 \rightarrow \frac{\sigma_X^2}{n}$$

Both of these assertions play a role in explicating the rationale for why

$$s_x^2 = \frac{\sum(x - \bar{x})^2}{n-1} \longrightarrow \sigma_x^2$$

To understand this rationale consider the fact that if

$$\frac{\sum(x - \mu_x)^2}{n}$$

is an unbiased estimate of σ_x^2, the quantity

$$\frac{\sum(x - \bar{x} + \bar{x} - \mu_x)^2}{n}$$

must also be an unbiased estimate of σ_x^2. This is because the expression $\sum(x - \mu_x)^2$ is algebraically equal to $\sum(x - \bar{x} + \bar{x} - \mu_x)^2$

since all we did was to subtract and then add the same value ($\bar{X}$ = the mean of the sample) to each $(x - \mu_x)$

but $$\sum(x - \bar{x} + \bar{x} - \mu_x)^2 =$$

$$\sum[(x - \bar{x}) + (\bar{x} - \mu_x)]^2 =$$

$$\sum[(x - \bar{x})^2 + 2(x - \bar{x})(\bar{x} - \mu_x) + (\bar{x} - \mu_x)^2]$$

To see why this is so we should recall that in high school algebra we learned that

$$(a + b)^2 = a^2 + 2ab + b^2$$

$$\begin{array}{r} (a + b) \\ \times\,(a + b) \\ \hline ab + b^2 \\ a^2 + ab \quad\; \\ \hline a^2 + 2ab + b^2 \end{array}$$

i.e., we multiply these quantities in the same fashion as we would multiply the numbers

$$\begin{array}{r} (5 + 3) \\ \times\,(5 + 3) \\ \hline 15 + 9 \\ 25 + 15 \quad\; \\ \hline 25 + 30 + 9 = \boxed{64} \end{array} \qquad \text{Just as} \qquad \begin{array}{r} 8 \\ \times\, 8 \\ \hline 64 \end{array}$$

If we think of $(X - \bar{X})$ as being (a) and $(\bar{X} - \mu_x)$ as being (b) we can see how

$$[(X - \bar{X}) + (\bar{X} - \mu_x)]^2 =$$

$$[(X - \bar{X})^2 + 2(X - \bar{X})(\bar{X} - \mu_x) + (\bar{X} - \mu_x)^2]$$

We also know from what we have learned about using the summation sign that

$$\sum\left[(X-\bar{X})^2 + 2(X-\bar{X})(\bar{X}-\mu_X) + (\bar{X}-\mu_X)^2\right]$$

$$= \sum(X-\bar{X})^2 + 2\left[\sum(X-\bar{X})\right]\left[(\bar{X}-\mu_X)\right] + n(\bar{X}-\mu_X)^2$$

Notice, however, that the middle term in this expression drops out because, as we have learned earlier, for any set of Xs, the $\sum(X-\bar{X})=0$. In other words, for a given sample of size n

$$\sum(X-\mu_X)^2 = \sum(X-\bar{X})^2 + n(\bar{X}-\mu_X)^2$$

This means that for a given sample of size n the statistic

$$\frac{\sum(X-\mu_X)^2}{n} = \frac{\sum(X-\bar{X})^2}{n} + (\bar{X}-\mu_X)^2$$

From this we can deduce that since (as was demonstrated earlier) the statistic $\frac{\sum(X-\mu_X)^2}{n}$ is an unbiased estimate of σ_X^2 and since (as was also demonstrated earlier) the statistic $(\bar{X}-\mu_X)^2$ is an unbiased estimate of $\frac{\sigma_X^2}{n}$ the statistic $\frac{\sum(X-\bar{X})^2}{n}$ must be an unbiased estimate of $\sigma_X^2 - \frac{\sigma_X^2}{n}$.

Let's now look at the quantity $\sigma_X^2 - \frac{\sigma_X^2}{n}$

Algebraically

$$\sigma_X^2 - \frac{\sigma_X^2}{n} = \frac{n}{n}\sigma_X^2 - \frac{\sigma_X^2}{n}$$

$$= \frac{n\sigma_X^2 - \sigma_X^2}{n}$$

$$= \frac{(n-1)\,\sigma_X^2}{n}$$

$$= \frac{n-1}{n}\sigma_X^2$$

This means that

$$\frac{\sum(X-\bar{X})^2}{n} \longrightarrow \frac{(n-1)}{n}\sigma_X^2$$

Another way to say the same thing is to assert that the mean of the sampling distribution of the statistic

$$\frac{\sum(X-\bar{X})^2}{n} \quad \text{is} \quad \frac{(n-1)}{n}\sigma_X^2$$

We know, however (from the transformation principles that we studied earlier), that if we multiply every item in a sampling distribution by a constant, we multiply the mean of that sampling distribution by that constant.

This means that

$$\left[\frac{n}{n-1}\right]\left[\frac{\sum(X-\bar{X})^2}{n}\right] \rightarrow \left[\frac{n}{n-1}\right]\left[\frac{(n-1)}{n}\right]\sigma_X^2$$

In other words, if we carry out the multiplication on both sides of the above expression we discover that

$$\boxed{s_x^2 = \frac{\sum(x-\bar{x})^2}{n-1} \longrightarrow \sigma_x^2}$$

This, of course, is what we set out to show. Admittedly, the demonstration has been lengthy and complicated, but for the reasons noted in the early part of this chapter, this could not be avoided. It may be helpful, therefore, to try to summarize what we have just done. In essence, our analytic procedure involved five steps.

1) We showed that (A) the statistic $\dfrac{\sum(x-\mu_x)^2}{n}$

computed on a random sample of n items is an unbiased estimate of σ_x^2

Our strategy was to show that $\dfrac{\sum(x-\mu_x)^2}{n}$

can be regarded as the mean of a random sample of n items from a population of

$(x - \mu_x)^2$s. Since we can show that

$\mu(x-\mu_x)^2 = \sigma_x^2$ statement A must be true.

2) We showed that (B) the statistic $(\bar{x}-\mu_x)^2$ computed on the same random sample of n items is an unbiased estimate of $\dfrac{\sigma_x^2}{n}$

Our strategy was to show that $(\bar{x}-\mu_x)^2$ can be regarded as a randomly selected item from a population of $(\bar{x}-\mu_x)^2$s. Since we can show that $\mu(\bar{x}-\mu_x)^2 = \sigma_{\bar{x}}^2 = \dfrac{\sigma_x^2}{n}$ statement B must be true.

3) We showed that $\dfrac{\sum(X - \mu_X)^2}{n}$ computed on a given sample is algebraically equal to

$$\frac{\sum(X - \bar{X})^2}{n} + (\bar{X} - \mu_X)^2$$

4) We showed that since

$$\frac{\sum(X - \mu_X)^2}{n} \rightarrow \sigma_X^2$$

and since

$$(\bar{X} - \mu_X)^2 \rightarrow \frac{\sigma_X^2}{n}$$

the quantity

$$\frac{\sum(X - \bar{X})^2}{n} \rightarrow \sigma_X^2 - \frac{\sigma_X^2}{n}$$

5) We used several algebraic transformations to show that if

$$\frac{\sum(X - \bar{X})^2}{n} \rightarrow \sigma_X^2 - \frac{\sigma_X^2}{n}$$

it follows that

$$s_X^2 = \frac{\sum(X - \bar{X})^2}{n - 1} \rightarrow \sigma_X^2$$

We have just gone through a rather complex account of why s_X^2 is the appropriate statistic to use when we want to estimate σ_X^2 from the information in a random sample. We can gain further insight into this rationale if we consider that when we calculate s_X^2 from the information in a sample we do so because we do not know the value of μ_X (the mean of the population from which the sample was drawn). If we did have this information we would be able to use the statistic $\frac{\sum(X-\mu_x)^2}{n}$ to estimate σ_x^2.

When we do not know μ_x and we want to estimate σ_x^2 we must first calculate the $\bar{X}$ of the sample (to estimate μ_x) and then we proceed to calculate the several (n) squared deviations from the obtained $\bar{X}$.

We did not mention it earlier, but among its important properties, the mean of a set of numbers is the single value that will minimize the sum of squared deviations from itself. Accordingly, when we calculate $\sum(X-\bar{X})^2$ on a given set of numbers the sum we obtain will be as small as it possibly could be for those numbers. Consider, for example, the set of three numbers 3, 4, and 11

the mean is $\frac{3+4+11}{3} = \frac{18}{3} = 6$

and
$$(3-6)^2 = (-3)^2 = 9$$
$$(4-6)^2 = (-2)^2 = 4$$
$$(11-6)^2 = (+5)^2 = 25$$

In other words, the sum of squared deviations from the mean of this set of numbers is 38.

Now let's take the sum of squared deviations from some value that is not the mean of this set. Let's try a number that is less than 6.

$$(3-5)^2 = (-2)^2 = 4$$
$$(4-5)^2 = (-1)^2 = 1$$
$$(11-5)^2 = (+6)^2 = 36$$

As can be seen, the sum of the squared deviations from 5 is 41 which is larger than 38.

Let's try again, but this time we'll take our deviations around a value that is greater than 6.

$$(3-7)^2 = (-4)^2 = 16$$

$$(4-7)^2 = (-3)^2 = 9$$

$$(11-7)^2 = (+4)^2 = 16$$

$$\text{Sum} = 41$$

Clearly this is no better.

While we could go on trying new numbers ad infinitum we can avoid much work if we analyze the issue using some algebra.

Here is that analysis. Let's begin with the notion that the quantity ($\bar{X} + C$) will be some number other than the mean of the set (except when C = 0). When expressed in this fashion our problem is to find that value of C that permits the expression

$\sum[x - (\bar{x}+c)]^2$ to be minimized

$$\sum[x - (\bar{x}+c)]^2 = \sum[(x-\bar{x}) + c]^2$$

$$= \sum[(x-\bar{x})^2 + 2c(x-\bar{x}) + c^2]$$

$$= \sum(x-\bar{x})^2 + 2c\sum(x-\bar{x}) + nc^2$$

$$\text{but } \sum(x-\bar{x}) = 0 \quad \therefore$$

$$\sum[x - (\bar{x}+c)]^2 = \sum(x-\bar{x})^2 + nc^2$$

When expressed in this fashion, it is clear that the expression will have minimal value when and only when C = 0. All of this means, as stated earlier:

The mean of a set of items is the one value that minimizes the sum of the square deviation from itself.

This property of the mean has important implications when our estimate of σ_x^2 must be based only on the information in a sample.

Consider the two statistics

$$\frac{\sum(x-\bar{x})^2}{n} \text{ and } \frac{\sum(x-\mu_x)^2}{n}$$

We showed earlier that $\frac{\sum(x-\mu_x)^2}{n}$ is an unbiased estimate of σ_x^2. We also showed that $\frac{\sum(x-\bar{x})^2}{n}$ is an unbiased estimate of $\sigma_x^2 - \frac{\sigma_x^2}{n}$

The fact that the mean is the one number that minimizes the sum of squared deviations in a given set of numbers helps to explain why the latter fact is so.

Suppose that we have a particular sample of n items. If we knew the value of μ_x we could calculate $\frac{\sum(x-\mu_x)^2}{n}$ for that sample and produce an unbiased estimate of σ_x^2

Suppose, on the other hand, we have the same set of items and we do not know μ_x . In this case we would use the mean of the set of items to estimate and we would then calculate the sum of the squared deviations from that estimate, i.e., we would calulate $\sum(x-\bar{x})^2$. In doing so, however, we would be deriving the smallest possible sum of squares that the particular set of items in the sample could produce.

Again compare $\frac{\sum(x-\bar{x})^2}{n}$ and $\frac{\sum(x-\mu_x)^2}{n}$

Since, in most instances, the mean of a given sample will not exactly equal μ_X, for most samples the statistic $\sum(X - \bar{X})^2$ will tend to be smaller than the statistic $\sum(X - \mu_X)^2$ calculated on the same sample. In other words, if the statistic $\frac{\Sigma(X - \mu_X)^2}{n}$ is an unbiased estimate of σ_X^2 the statistic $\frac{\Sigma(X - \bar{X})^2}{n}$ must tend to underestimate σ_X^2. Indeed, across a large number of samples it ought to underestimate σ_X^2 by an amount that is determined by the variability of means of samples of size n (i.e., by the differences between σ_X^2 and $\frac{\sigma_X^2}{n}$). This is exactly what was expressed when it was asserted earlier that

$$\frac{\Sigma(X - \bar{X})^2}{n} \longrightarrow \sigma_X^2 - \frac{\sigma_X^2}{n}$$

These considerations tell us why, when we use only the information in a sample to estimated σ_X^2, we must divide $\sum(X - \bar{X})^2$ by a number that is less than n. In essence, we do so to make our estimate a little larger and thereby correct for the bias that would otherwise occur.

Let's now examine the two statistics:

$$\frac{\sum(X - \bar{X})^2}{n-1} \quad \text{and} \quad \frac{\sum(X - \mu_X)^2}{n}$$

Both of them are unbiased estimates of σ_X^2 (i.e., the mean of the sampling distribution of both statistics is σ_X^2), but they are not equally good estimates because for a given sample size (n) the variance of the sampling distribution of $\frac{\Sigma(X - \bar{X})^2}{n-1}$ will always be somewhat larger than the variance of the sampling distribution of $\frac{\sum(X - \mu_X)^2}{n}$. This is because in estimating σ_X^2 with the statistic $\frac{\sum(X - \bar{X})^2}{n - 1}$, we in effect are using only (n - 1) of the obtained deviations from the mean to form our estimate and from the Central Limit Theorem we would expect that a statistic based on n - 1 items would tend to be more variable than a statistic based on n items. To see how this comes about, consider the situation that arises when we have a given set of Xs before us, i.e., we have a particular random sample of size n. If we know the value of μ_X we subtract it from each X and square each of the resulting set of values. This gives us a set of n randomly selected squared deviations from μ_X each of which might be any value whatsoever. Now, consider what happens when we must use the information in the sample to estimate μ_X. First we calculate $\bar{X} = \frac{\sum X}{n}$ and then we square each of the deviations from $\bar{X}$. This also gives us a set of numbers on which we will base our estimate of σ_X^2, but this time only n - 1 of the numbers are free to assume any value whatsoever. This is because we used the $\bar{X}$ of the set of items in the sample in calculating each of the square deviations and, as we learned earlier, for a given set of items the sum of the deviations from the mean of the set is always zero. This implies that once we have carried our calculations to the point where we have specified n - 1 of the squared deviations we can always specify the value of the last entry without actually looking at it. The last entry in our calculations must always be such that the value of $\bar{X}$ it entails will make the sum of the deviations from $\bar{X}$ equal zero. In other words, because we have used the mean of our sample to estimate σ_X^2 we have imposed a constraint on the values that make up our set of squared deviations and these are the randomly selected values that we use to estimate σ_X^2. Statisticians describe this set of circumstances by asserting that the statistic $\frac{\sum(X - \mu_X)^2}{n}$ has n degrees of freedom, whereas the statistic $\frac{\Sigma(x - \bar{x})^2}{n - 1}$ has only n - 1 degrees of freedom.

As we proceed to examine various other statistics we will often find ourselves having to specify the number of degrees of freedom that characterizes a given statistic. In general, we will find that the number of degrees of freedom for a given statistic will be equal to the number of values in our final calculations that are, in fact, free to vary and that whenever we use the information in our sample to estimate a parameter in the course of obtaining those values, we will lose a degree of freedom.

Chapter 6

SAMPLING DISTRIBUTIONS OF STATISTICS THAT ARE BASED ON s_x^2

We have just seen why the mean of the sampling distribution of s_x^2 is equal to σ_x^2. With a certain amount of thought, it should become apparent that unlike $\bar{X}$ or $z_{\bar{X}}$ which are distributed normally when the samples come from a normal population, the sampling distribution of s_x^2 will be skewed even when the samples come from a population that is normal. We can see why, and in the course of doing so gain some insight into the sampling distribution of s_x^2 if we look carefully at the sampling distribution of

$$\frac{\sum(X-\mu_x)^2}{n}$$

. We saw in the previous chapter that like s_x^2, the sampling distribution of the statistic $\frac{\sum(X-\mu_x)^2}{n}$ has a mean equal to σ_x^2. Let's examine that sampling distribution in some detail.

Suppose that we begin with a quite crude approximation to a normal population of Xs with a given mean and variance.

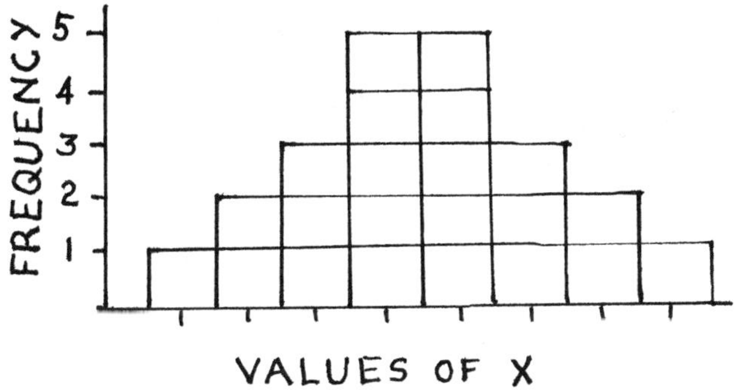

Suppose that we subtract μ_x from every item in the above population and then form a new population of N items where each item consists of $|X-\mu_x|$.

Here is that population.

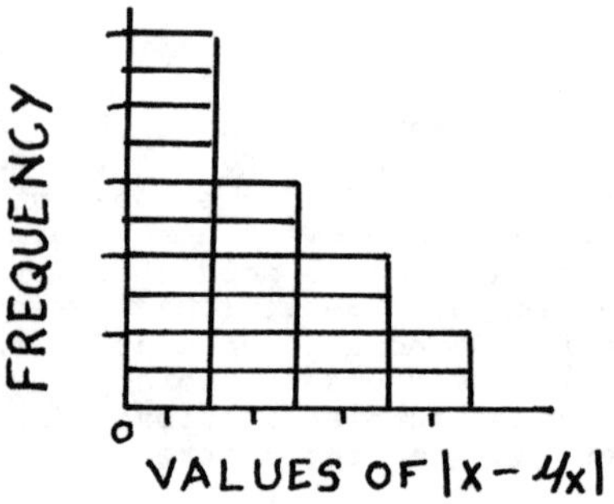

Notice that there are no negative values of $|X - \mu_X|$ and that small values of $|X - \mu_X|$ are much more frequent than large values.

Suppose we square each $|X - \mu_X|$ in this population. When we do so we get a new population of N numbers in which small values of $|X - \mu_X|^2$ are again much more frequent than large values. Moreover, since $|X - \mu_X|^2 = (X - \mu_X)^2$ the mean of this new set of numbers $\mu_{(X-\mu_X)^2} = \frac{\sum (X - \mu_X)^2}{N}$ must be the variance of the original set set of Xs from which these numbers were derived.

If we conceive of the statistic $\frac{\sum (X - \mu_X)^2}{n}$ as the mean of a sample of n items from the population of $(X - \mu_X)^2$s, we know (from the Central Limit Theorem) that the mean of the sampling distribution of $\frac{\sum (X - \mu_X)^2}{n}$ will equal the mean of the population from which the $(X - \mu_X)^2$s were drawn.

$$\mu\left[\frac{\sum (X - \mu_X)^2}{n}\right] = \sigma_X^2$$

We also know (from part 3 of the Central Limit Theorem) that even though the population of $(X - \mu_X)^2$ is highly skewed, as sample size increases (i.e., increases) the sampling distribuiton of $\frac{\sum(X - \mu_X)^2}{n}$ will begin to lose its skew and increasingly approximate a normal curve.

All of this suggests that even when our samples come from a normal population, the statistic $\frac{\sum (X - \mu_X)^2}{n}$ will have a sampling distribution that is very skewed for small samples and it will only gradually tend to lose its skew as n increases.

With some thought it should become apparent that although we have been dealing with the statistic $\frac{\sum (X - \mu_X)^2}{n}$ these same arguments can help us conceptualize the shape of the sampling distribution of

$$s_X^2 = \frac{\sum (X - \bar{X})^2}{n - 1}$$

Thus, if we repeatedly took random samples of size n = 7 from a population that was normal with $\sigma_X^2 = 4$ and we calculated $s_X^2 = \frac{(X - \bar{X})^2}{n - 1}$ on each sample, the sampling distribution of s_X^2 that would eventually be obtained should have the following form.

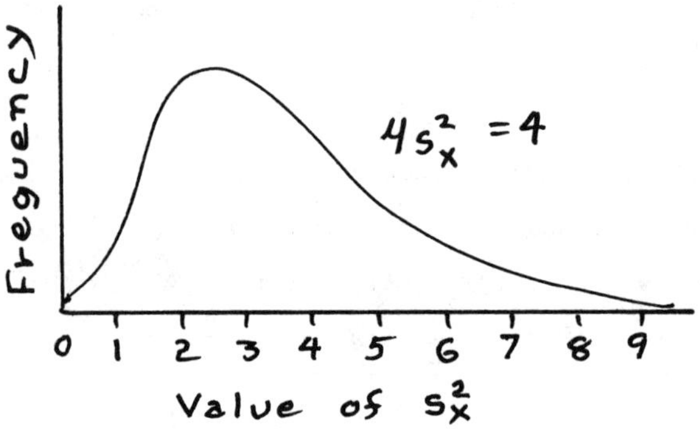

Since n = 7 and since the degrees of freedom = (n - 1) the df for the s_x^2 in this distribution is 6.

We should also be able to see that the shape of the sampling distribution of s_x^2 will be determined by the size of the samples upon which it is based or, properly speaking, that its shape will depend upon the degrees of freedom in s_x^2 .

Here are some of the sampling distributions of s_x^2 that are obtained when the samples come from a normal population with $\sigma_x^2 = 4$

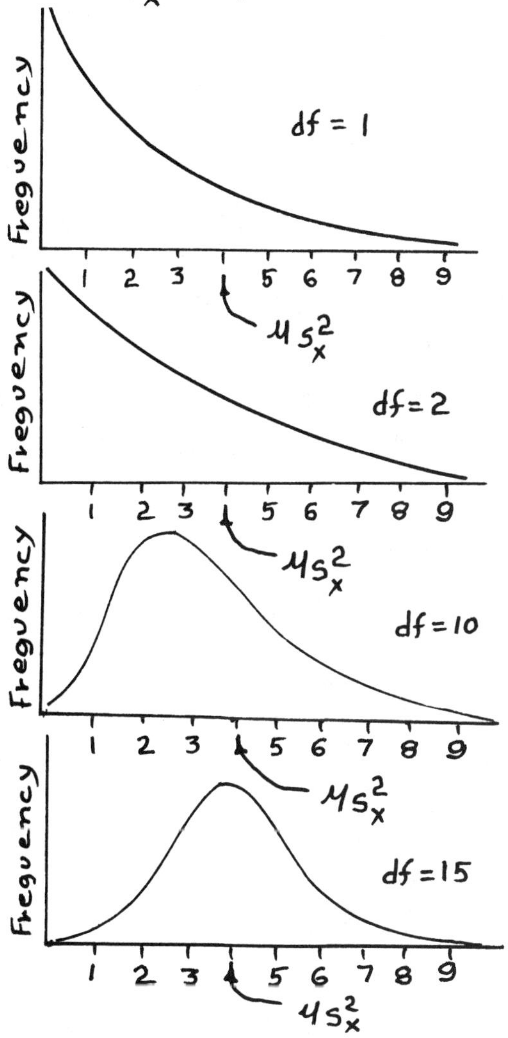

As can be seen, and as we would expect from what we now know about sampling, all of the distributions have a mean that is equal to σ_X^2 and as the sample size (and accordingly df) increases, the sampling distribution of s_X^2 begins to exhibit less and less skew. Moreover, as might also be expected, as df increases, the variance of the sampling distribution of s_X^2 gets smaller and smaller. Of course, it has to because the larger our samples, the more closely will the value of each s_X^2 approximate σ_X^2.

The Sampling Distribution of $\frac{s_X^2}{\sigma_X^2}$

We can transform a given sampling distribution of s_X^2 into a new distribution with a mean of 1 if we divide every s_X^2 in it by σ_X^2 (the variance of the population from which the samples were drawn). Another way to say the same thing is to assert that the mean of the sampling distribution of the statistic $\frac{s_X^2}{\sigma_X^2}$ is equal to one when the samples come from a population with variance equal to σ_X^2. Notice that we do not have to know the value of μ_X (the mean of the population from which the samples were drawn) to form a given sampling distribution of $\frac{s_X^2}{\sigma_X^2}$. We do, on the other hand, have to know the value of σ_X^2, if that sampling distribution is to in fact have a mean of one. When the samples come from a normal population with variance equal to σ_X^2 the statistic $\frac{s_X^2}{\sigma_X^2}$ is distributed as follows:

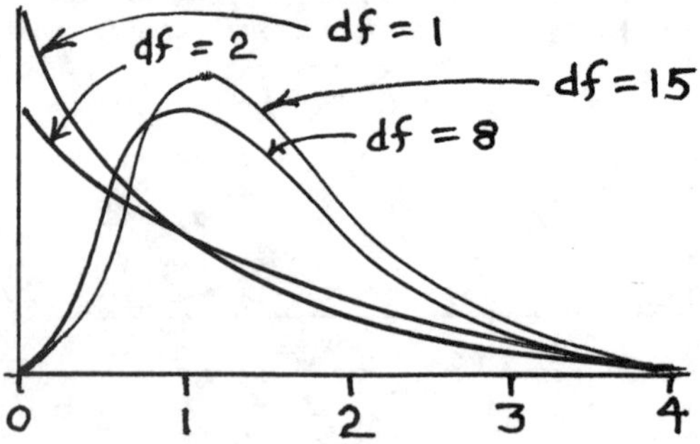

Of course, there are an infinite possible number of sampling distributions of $\frac{s_x^2}{\sigma_x^2}$, because like the sampling distribution of s_x^2 , the shape of the sampling distribution of $\frac{s_x^2}{\sigma_x^2}$ will be determined by sample size. Despite this complication $\frac{s_x^2}{\sigma_x^2}$ is such a useful statistic that statisticians have calculated the values in the sampling distribution of $\frac{s_x^2}{\sigma_x^2}$ that cut off certain selected proportions of the total area for a reasonably large variety of possible sample sizes.

For example, the first two columns of Table II in the appendix show (for a variety of df's), the values of $\frac{s_x^2}{\sigma_x^2}$ that cut off the lower and upper .005 of the sampling distribution that is obtained when the samples come from a normal population with variance equal to σ_x^2. The second two columns in Table II show the values of $\frac{s_x^2}{\sigma_x^2}$ that cut off the lower and upper .025 of these distribuitons. This information is valuable because it provides us with a way of testing a statistical hypothesis about σ_x^2. Our procedure would be essentially the same as when we test a hypothesis about μ_x . This time, however, we draw a random sample and calculate $\frac{s_x^2}{\sigma_h^2}$ where

$$s_x^2 = \frac{\sum(x-\bar{x})^2}{n-1}$$ calculated on the sample and σ_h^2 is the hypothesized value of σ_x^2 .

From the material we have just discussed we know that if the hypothesis tested is true (i.e. $\sigma_h^2 = \sigma_x^2$) the sampling distribution of the statistic $\frac{s_x^2}{\sigma_h^2}$ will have a mean of one and that Table II lists the values of $\frac{s_x^2}{\sigma_h^2}$ that cut off the indicated extremes of the distribution. If our obtained value of $\frac{s_x^2}{\sigma_h^2}$ falls between those extremes we have no basis to reject the hypothesis tested. If, on the other hand, the obtained value of $\frac{s_x^2}{\sigma_h^2}$ falls in one of the rejection regions we can reject the hypothesis that $\sigma_x^2 = \sigma_h^2$ knowing that we do so under conditions where the probability of rejecting a true hypothesis is exactly equal to the probability of obtaining values of $\frac{s_x^2}{\sigma_h^2}$ that are either smaller or larger than the critical values listed in Table II.

Often, (for reasons that will be explained shortly) we are only concerned with controlling the probability of rejecting the hypothesis that $\sigma_x^2 = \sigma_h^2$ when we have reason to suspect that σ_x^2 may, in fact, be greater than σ_h^2. Under such circumstances it would not be to our advantage to employ a testing strategy in which we would sometimes reject the hypothesis tested when σ_x^2 was less than σ_h^2. For example, according to the procedures outlined above, if the obtained value of $\frac{s_x^2}{\sigma_h^2}$ is smaller than the critical value listed in the first column of Table II we would reject the hypothesis tested even though the information in our sample (i.e., the value s_x^2) leads us to suspect that σ_x^2 might be smaller (rather than larger) than σ_h^2.

If our concern was in controlling the probability of rejecting the hypothesis tested only when we had reason to suspect that σ_x^2 exceeded σ_h^2, we would employ a slightly different strategy. Instead of dividing our rejection region between the two tails of the sampling distribution we would only reject the hypothesis tested when the obtained value of $\frac{s_x^2}{\sigma_h^2}$ was among the extremely large values that occur with probability equal to α when the hypothesis tested is true. Another way to say the same thing is to assert that we would carry out a one-tailed test in which the entire rejection region was assigned to the right hand side of the sampling distribution of $\frac{s_x^2}{\sigma_h^2}$

The fifth column in Table II shows, for a given df, the value of $\frac{s_x^2}{\sigma_h^2}$ that cuts off the upper .05 of the sampling distribution of $\frac{s_x^2}{\sigma_h^2}$ that is obtained when the samples come from a normal population and $\sigma_x^2 = \sigma_h^2$. The sixth column in Table II provides the critical values of $\frac{s_x^2}{\sigma_h^2}$ that cut off the upper .01 of these distributions.

Again, a practical example will help to clarify the manner in which we use this information. Let's once more imagine that a manufacturer of light bulbs has equipment that is designed to produce bulbs with a mean burning time of 740 hours and that he knows that when the equipment is working according to specifications the S.D. of the burning times of the bulbs it produces is 20 hours (i.e., $\sigma^2_{\text{burning time}} = 400$.

In Chapter 3 we saw how the manufacturer could decide if his equipment is turning out bulbs that on the average had the appropriate burning life. To do so, he set up the statisitical hypothesis: $\mu_x = \mu_h = 740$ and he proceeded to use the statistic

$$z_{\bar{x}} = \frac{\bar{X} - \mu_h}{\sqrt{\frac{\sigma_x^2}{n}}}$$

to test his hypothesis. This was a reasonable

procedure because he knew that if the equipment went out of adjustment the average burning times of the bulbs it produced might increase or decrease but the variance among the bulbs would not ordinarily change.

There are, however, certain circumstances in which the variability among bulbs might conceivably increase over time. For instance, one can imagine that if the bearings on the equipment became worn, either through extended use or through improper maintenance, the variability among bulbs might well increase. Moreover, one can conceive of this happening even though the equipment remained in adjustment and hence continued to turn out bulbs with an average burning life of 740 hours. The procedures we have been studying can help the manufacturer decide whether or not the bearings on this equipment have become worn and hence, introduced unwanted variability into his product.

As before, he would draw a random sample of bulbs, but this time he would calculate

$s_x^2 = \frac{\sum(X-\bar{X})^2}{n-1}$ on his sample. He would then form the statistic $\frac{s_x^2}{\sigma_h^2}$ by dividing the obtained value of s_x^2 by the value of σ_x^2 (i.e., 400) that he hypothesizes to characterize the population from which the sample was drawn. In other words, he would hypothesize that the equipment was functioning properly and that the variance of its output was 400 hours. Let's suppose that he has decided to test his hypothesis under conditions where the probability of rejecting it, if it is true, is exactly .05. Under these circumstances his next step would be to ask if the obtained value of $\frac{s_x^2}{\sigma_h^2}$ was among those large values of $\frac{s_x^2}{\sigma_h^2}$ that would only occur .05 of the time when samples of size n (df = n - 1) are drawn from a normal population with σ_x^2 = 400. Column 5 of Table II would give him the critical value of $\frac{s_x^2}{\sigma_h^2}$

that would lead him to reject the hypothesis tested at the α = .05 level of confidence. Column 6 of Table II would give him the critical value that he would need to reject the hypothesis if he had elected to set α = .01 instead of α = .05.

Notice that with these procedures a value of s_x^2 that was less than 400 and accordingly, a value of

$\frac{s_x^2}{\sigma_h^2} < 1$ would provide no basis for suspecting

that the variance among bulbs was too large. This, of course, is as it should be. After all, in this instance the manufacturer is trying to detect and repair bearing wear and it would be unreasonable to tear down his machines when a given sample suggested that the equipment might be functioning better than ordinarily expected.

This leads to our second example of a statistical test

Example II - Testing a Statistical Hypothesis about σ_x^2

Illustration

In order to determine if certain equipment is producing light bulbs that are too variable (the variance of their burning times is larger than desirable) we test the hypothesis that σ_x^2 is equal to the variance of the burning times indicated in the original equipment specifications.

Method

(1) Hypothesis: Population variance is equal to original specification, i.e., σ_x^2 = a specified value = σ_h^2

(2) Set α

(3) Use $$\frac{s_x^2}{\sigma_h^2} = \frac{\frac{\sum(x-\bar{x})^2}{n-1}}{\sigma_h^2}$$

where n = Sample size

$\bar{X}$ = sample mean

σ_h^2 = hypothesized population variance

(4)

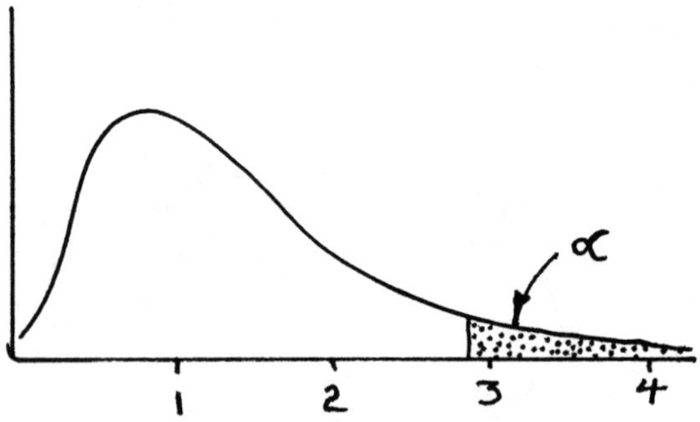

We determine the sampling distribution of $\frac{s_x^2}{\sigma_h^2}$ under the assumption that the hypothesis is true and we specify the location of the rejection region.

Because we are only interested in rejecting the hypothesis when we have evidence that suggests that $\sigma_x^2 > \sigma_h^2$ we use a one-tailed test.

(5) Compute $\frac{s_x^2}{\sigma_h^2}$ and determine whether or not it falls in the rejection region.

Notice in this illustration that if we can reject the hypothesis tested we will have reason to believe that the variance of the equipment's output is larger than desirable.

The Sampling Distribution of F

Another important statistic that is based on s_x^2 is the F ratio (named for the great statistician R. A. Fisher).

To calculate a given F ratio we would draw two random samples of size n_1 and n_2 from a given normal population and calculate s_x^2 on each of them. We would then form the ratio:

$$F_{df=(n_1-1),(n_2-1)} = \frac{s_{x_1}^2}{s_{x_2}^2} = \frac{\dfrac{\Sigma(X_1-\bar{X}_1)^2}{n_1-1}}{\dfrac{\Sigma(X_2-\bar{X}_2)^2}{n_2-1}}$$

We saw earlier that the nature of the sampling distribution of s_x^2 was determined by the degrees of freedom associated with s_x^2. The nature of the sampling distribution of F is determined by the degrees of freedom associated with each of the two s_x^2s that make up F. This means that every possible pair of sample sizes will yield a different sampling distribution of F. Here, for example, are 3 of the many possible distributions of F.

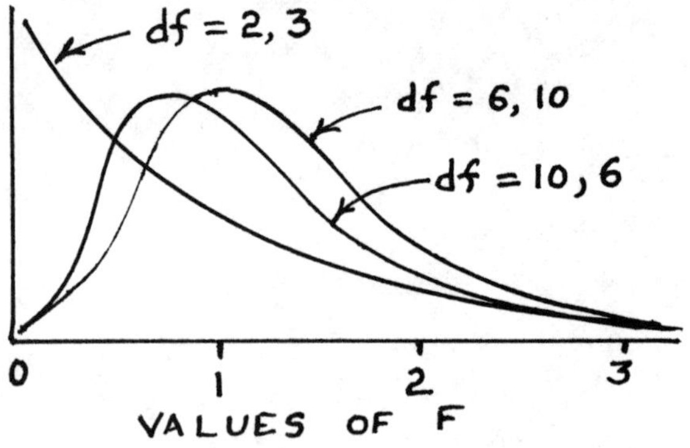

By convention, the first number refers to the df forthe numerator of F. Notice that the sampling distribution of $F_{df = 6, 10}$ is not the same as the sampling distribution of $F_{df = 10, 6}$

The sampling distribution of F has numerous practical applications and moreover, as we will discover, many other sampling distributions turn out to be special cases of the F distribution. For example, a moment's thought should make it clear that the sampling distribution of F with df = $(8, \infty)$ is the sampling distribution of $\frac{s_X^2}{\sigma_X^2}$ for samples of size 9.

This follows because if we took pairs of samples such that the first always consisted of 9 items and the second was infinitely large, the second s_X^2 would always exactly equal σ_X^2. Accordingly, our obtained value of $\frac{s_{X_1}^2}{s_{X_2}^2}$ would always be exactly $\frac{s_{X_1}^2}{\sigma_X^2}$

Like the sampling distribution of $\frac{s_X^2}{\sigma_X^2}$, the sampling distribution of F = $\frac{s_{X_1}^2}{s_{X_2}^2}$ has a mean equal to one. Moreover, like the sampling distribution of $\frac{s_X^2}{\sigma_X^2}$, as the df in the numerator and denominator of F increases, the sampling distribution of F loses some of its skew and its variance decreases.

An obvious application of the F ratio occurs whenever we wish to test the statistical hypothesis that two random samples have been selected from the same normal population.

Since we would obtain the same sampling distribution of F when our samples are drawn from two normal populations with $\sigma_{X_1}^2 = \sigma_{X_2}^2$ as when both samples come from a single normal population (note that the means of the two populations do not have to be equal) we can also use F to test the statistical hypothesis that a given pair of random samples come from populations with $\sigma_{X_1}^2 = \sigma_{X_2}^2$

Another way to say essentially the same thing is to assert that we can use the F ratio to decide whether or not the s_X^2 computed on each of two samples are both estimating the same σ_X^2.

Table III in the appendix shows the values of F that cut off the upper .05 of the sampling distribution of F that is obtained when both of its component samples come from normal populations with

$\sigma_{X_1}^2 = \sigma_{X_2}^2$

Each entry in the body of Table III is the critical value for a different F distribution. More specifically, a given entry is the critical value of F for the distribution obtained when the s_X^2 in the numerator of F has the df indicated at the head of the column and the s_X^2 in the denominator has the df indicated at the head of the row in which the entry appears. Each entry in Table IV is the critical value of F that cuts off the upper .01 of the F distribution with df's indicated at the head of the appropriate row and column.

There are a number of subtle features about the tabled values of F that may not be apparent on first glance but which, nevertheless, deserve our careful attention.

First of all it is important to emphasize that the entries in Tables III and IV are the critical values that cut off the extremely large values of F that occur 5% of the time (For Table III) or 1% of the time (For Table IV). Thus, for example, when we look in Table III we find that when the numerator of F has 2 df and its denominator has 6 df the critical values of F is 5.14. This means that for the F distribution with 2, 6 df the value of F that cuts off the upper 5% is 5.14.

Here is the sampling distribution of F with df = 2, 6.

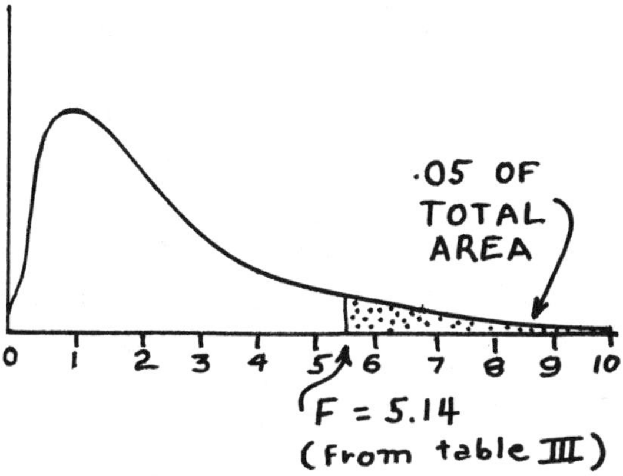

Here, again, is the sampling distribution of F df = 2, 6 but this time the value of F that cuts off the upper 1% of the distribution has been indicated.

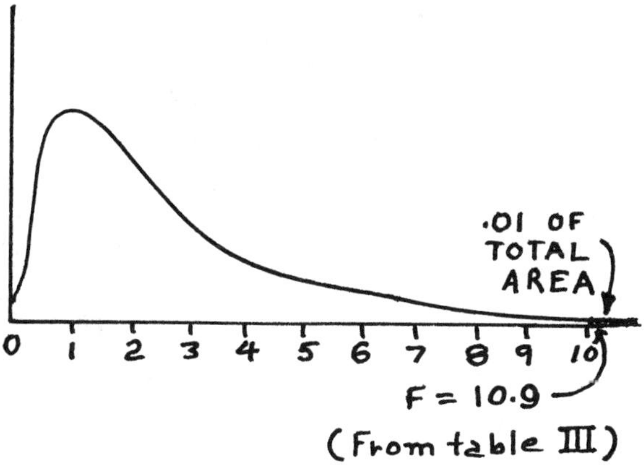

As seen in these examples the entries in Tables III and IV are critical values for a one-tailed test. More specifically, these are the values that specify the rejection region for $\alpha = .05$ and $\alpha = .01$ when

1) we wish to test the hypothesis that $\sigma^2_{X_1} = \sigma^2_{X_2}$

2) we are only concerned with detecting those conditions where $\sigma^2_{X_1} > \sigma^2_{X_2}$ and 3) we want to be certain that the probability of rejecting the hypothesis when it is true is exactly equal to α. Under such circumstances we would not want to interpret an obtained $\frac{s^2_{X_1}}{s^2_{X_2}}$ that was less than one (i.e., an obtained $s^2_{X_1} < s^2_{X_2}$) as evidence that $\sigma^2_{X_1} > \sigma^2_{X_2}$. Accordingly, we assign the entire rejection region to the upper end of the F distribution.

There is, of course, nothing about the distribution that says we could not use it to carry out a two-tailed test, but if we needed to conduct one we would probably want more information than is provided by Tables III and IV. As we have just seen these tables were designed to show the critical values for a one-tailed test, and in particular a one-tailed test in which the rejection region encompasses the values of F that would lead us to believe that $\sigma^2_{X_1} > \sigma^2_{X_2}$. As might be expected, there is a reason why Tables III and IV were designed in this fashion. It stems from the fact that they were intended for use in analysis of variance (a topic we shall examine in later chapters). When we carry out an analysis of variance (for reasons to be elaborated later) we are especially interested in the situation where we have reason tc suspect that the numerator of F might be estimating a larger variance than its denominator.

If we had no other tables available, we could, of course, use the information in Tables III and IV to carry out a two-tailed test of the hypothesis that $\sigma^2_{X_1} = \sigma^2_{X_2}$ but we would have to employ a special strategy. For example, suppose we wanted to test the hypothesis that two samples came from populations where $\sigma_{X_1}^2 = \sigma_{X_2}^2$ and we were just as concerned about detecting the situation where $\sigma_{X_1}^2 < \sigma_{X_2}^2$ as where $\sigma_{X_1}^2 > \sigma_{X_2}^2$. Suppose further, that we draw our two random samples and calculate $s_{X_1}^2$ and $s_{X_2}^2$.

If $s_{X_1}^2 > s_{X_2}^2$ either $F = \dfrac{s_{X_1}^2}{s_{X_2}^2}$ will fall in the critical region indicated by the appropriate entry in Table III or IV or the F will fall short of the critical region. But what would it mean if our obtained $s_{X_1}^2 < s_{X_2}^2$. Our tables were designed for one-tailed tests and, hence, they indicate the cutoff point for large values of $F = \dfrac{s_{X_1}^2}{s_{X_2}^2}$ that occur .05 of the time (Table III) or .01 of the time (Table IV) when $\sigma_{X_1}^2 = \sigma_{X_2}^2$. They tell us nothing about the probabilities associated with any of the small values of F that also occur when $\sigma_{X_1}^2 = \sigma_{X_2}^2$.

One way to resolve this problem is to calculate

$$F_{df = (n_2 - 1), (n_1 - 1)} = \frac{s_{X_2}^2}{s_{X_1}^2} \text{ when } s_{X_1}^2 < s_{X_2}^2$$

and to then ask whether the obtained F exceeds the critical value listed for an F with df = $(n_2 - 1), (n_1 - 1)$.

But notice that if we employ this strategy we are conducting our test in such a way that when $\sigma^2_{X_1}$ does in fact equal $\sigma^2_{X_2}$ the probability of obtaining a significant value of F (i.e., a value that falls in a rejection region) is exactly double the probability listed for the table that we use. For example, with the strategy outlined here we can obtain a significant value of F if either of two events occur.

1) $s^2_{X_1} > s^2_{X_2}$ and the obtained value of

$$F = \frac{s^2_{X_1}}{s^2_{X_2}}$$ falls beyond the

critical value listed for df numerator = $(n_1 - 1)$ denominator = $(n_2 - 1)$

Because of the way Table III is constructed we know that the probability of this happening when $\sigma^2_{X_1} = \sigma^2_{X_2}$ is .05.

2) We can also obtain a significant value of F if $s^2_{X_1} < s^2_{X_2}$ and the obtained value of

$$F = \frac{s^2_{X_2}}{s^2_{X_1}}$$ falls beyond the

critical value listed for df numerator = $(n_2 - 1)$ df denominator = $(n_1 - 1)$

Because of the way Table III is constructed we also know that the probability of this happening when $\sigma^2_{X_1} = \sigma^2_{X_2}$ is .05.

Since we are willing to reject the hypothesis

tested (i.e., $\sigma^2_{X_1} = \sigma^2_{X_2}$) under either of these two conditions the total chance of rejecting the hypothesis tested (if it is true) must be equal to the sum of the probabilities associated with each conditon. In this example, then, α = .05 + .05 = .10.

Another way to say the same thing is to assert that if Tables III and IV are used to carry out a two-tailed test of the hypothesis that $\sigma^2_{X_1} = \sigma^2_{X_2}$ the α level for our test must be either .10 (in which case we would use Table III) or .02 (in which case we would use Table IV).

Because tables showing the critical values of F that cut off the upper regions of the F distribution are more common than tables showing values that cut off the upper <u>and</u> lower regions, it will be useful to summarize the strategy for carrying out a two-tailed F test when our tables are designed for one-tailed tests.

Example III - A Two-Tailed F Test of the Statistical Hypothesis that $\sigma_{x_1}^2 = \sigma_{x_2}^2$

Illustration

In order to determine if two samples come from populations with different variances we test the statistical hypothesis that they come from populations with the same variance.

Method

(1) Hypothesis $\sigma_{x_1}^2 = \sigma_{x_2}^2$

(2) Set α Note: If Tables III and IV (in the appendix of this book) are used, α must either be .1 or .02.

(3) If $s_{x_1}^2 > s_{x_2}^2$ compute

$$F_{df=(n_1-1),(n_2-1)} = \frac{s_{x_1}^2}{s_{x_2}^2}$$

If $s_{x_1}^2 < s_{x_2}^2$ compute

$$F_{df=(n_2-1),(n_1-1)} = \frac{s_{x_2}^2}{s_{x_1}^2}$$

where

n_1 = Size of sample 1

$$s_{x_1}^2 = \frac{\sum(X-\bar{X}_1)^2}{n_1-1} \quad \text{for sample 1}$$

n_2 = Size of sample 2

$$s_{x_2}^2 = \frac{\sum(X-\bar{X}_2)^2}{n_2-1} \quad \text{for sample 2}$$

(4)

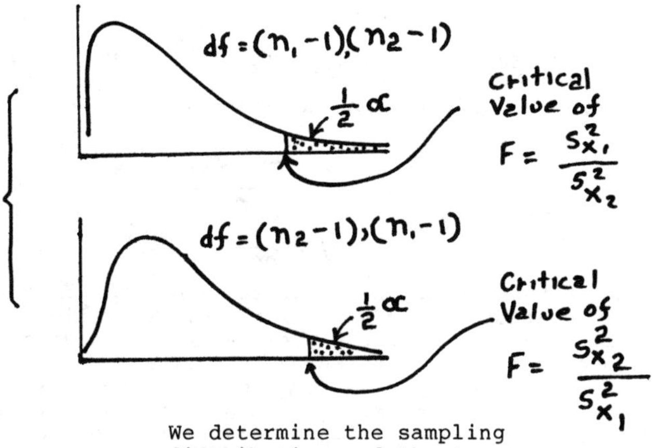

We determine the sampling distributions of the two possible Fs and we locate the 1/2 α rejection regions.

(5) Compute $s^2_{X_1}$ and $s^2_{X_2}$, form the appropriate F, and determine whether or not it falls in the appropriate rejection region.

Notice in this illustration that we have conducted our test in such a way that if the obtained F falls in one of the rejection regions we can reject the hypothesis tested at the α level of significance.

Chapter 7

TESTS OF HYPOTHESES ABOUT μ_X WHEN σ_X^2 IS NOT KNOWN

In Chapter 4 we saw how we could use the statistic

$$z_{\bar{X}} = \frac{\bar{X} - \mu_h}{\sqrt{\frac{\sigma_X^2}{n}}}$$

to test the hypothesis that a given random sample came from a population with μ_X equal to some specific value, i.e., μ_h . To carry out that test, however, it was necessary for us to known the value of σ_X^2 (the variance of the population from which the sample was drawn). We also learned (in Chapter 5) that if we do not know the value of σ_X^2 we can still test the hypothesis that $\mu_X = \mu_h$ if we use $\bar{X}$ calculated on our sample to estimate μ_X and it was asserted that under those circumstances we would use the statistic

$$t = \frac{\bar{X} - \mu_h}{\sqrt{\frac{s_X^2}{n}}}$$

There are certain obvious similarities between

$$z_{\bar{X}} = \frac{\bar{X} - \mu_h}{\sqrt{\frac{\sigma_X^2}{n}}} \quad \text{and} \quad t = \frac{\bar{X} - \mu_h}{\sqrt{\frac{s_X^2}{n}}}$$

The numerator of both statistics is the difference between the mean of the obtained sample and the hypothesized mean of the population from which the sample was drawn. However, where $z_{\bar{X}}$ expresses this difference in units of the known standard deviation of the sampling distribution of means, t expresses the difference in units that are an estimate of $\sigma_{\bar{X}}$. That is, where $\sqrt{\frac{\sigma_x^2}{n}} = \sigma_{\bar{X}}^2$ for samples of size n, the quantity $\sqrt{\frac{s_x^2}{n}} = s_{\bar{X}}^2$ is merely an estimate of $\sigma_{\bar{X}}^2$ based on the obtained sample of size n.

It will be recalled that when our samples came from a normal population with $\mu_X = \mu_h$ and variance = σ_X^2 the sampling distribution of $Z_{\bar{X}}$ is normal with mean = 0 and variance = 1. It is this fact that permits us to use a table of Z scores (Table 1) to assess the probability of obtaining various values of $Z_{\bar{X}}$ when our hypothesis is true.

Like $Z_{\bar{X}}$, when t is calculated on a sample from a normal population where $\mu_X = \mu_h$, the sampling distribution of t will have a mean of 0, but unlike $Z_{\bar{X}}$ the shape of the sampling distribution of t will depend upon sample size (n). When n is small the sampling distribution of t departs appreciably from a normal distribution and its variance is greater than one. Moreover, the smaller the samples, the greater the departure and the larger the variance of the t distribution. Another way to say this is to assert that like the sampling distributions of s_X^2, $\frac{s_x^2}{\sigma_X^2}$, and $\frac{s_{x_1}^2}{s_{x_2}^2}$, the shape of the sampling distribution of t will depend upon df and this means that if we are to specify a given distribution of t we must specify its df.

In situations where we use $t = \frac{\bar{X} - \mu_h}{\sqrt{\frac{s_x^2}{n}}}$ the df for t is (n - 1). We saw earlier (at the end of Chapter 5) that $s_x^2 = \frac{\Sigma(X - \bar{X})^2}{n-1}$ is conceived to contain n - 1 df because only (n - 1) of the n values of $(X - \bar{X})$ that make up a given s_X^2 are in fact, free to assume any values whatsoever. This is because the nth value of $(X - \bar{X})$ must always be such that the value of X, on which it is based will permit

$\sum(X - \bar{X})$ to equal zero. It was also asserted, at the time, that in general, we lose a degree of freedom whenever we must use the information in a sample to estimate a parameter in the course of calculating the value of a given statistic.

When we calculate a given value of $t = \dfrac{\bar{X} - \mu_h}{\sqrt{\dfrac{s_X^2}{n}}}$

we lose a degree of freedom because we must use the $\bar{X}$ of the sample to estimate μ_X in the course of calculating the value of the s_X^2 that appears in the demoninator of the t.

The next page shows a normal distribution with a sampling distribution of $t_{df=9}$ superimposed on it.

These are the sampling distributions of $z_{\bar{X}}$ and $t_{df=9}$ that are obtained when we take samples of

size n = 10 from a normal population where $\mu_X = \mu_h$ and on each sample 1) calculate $z_{\bar{X}}$, and also

calculate $t = \dfrac{\bar{X} - \mu_h}{\sqrt{s_X^2/n}}$. If we did so, both

distributions would have a mean of 0 since the samples would come from a population where $\mu_X = \mu_h$ but the variance of the sampling distribution of t would be larger than the variance of the sampling distribution

of $z_{\bar{X}}$, thus where $\sigma^2_{z_{\bar{X}}} = 1$ we would find that

$\sigma^2_{t_{df=9}} > 1$.

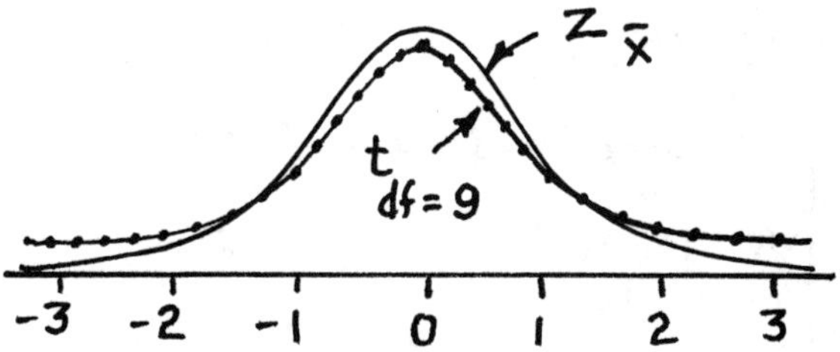

If we carefully examine these two distributions we will discover that the principal way in which the sampling distribution of $t_{df=9}$ departs from a normal distribution is that it contains too many large positive and large negative values.

If we again consider the formula for t and compare it to the formula for $Z_{\bar{X}}$ we can see why this happens.

$$Z_{\bar{X}} = \frac{\bar{X} - \mu_h}{\sqrt{\frac{\sigma_X^2}{n}}} \qquad t_{df=9} = \frac{\bar{X} - \mu_h}{\sqrt{\frac{S_X^2}{n}}}$$

As revealed here, the only difference between the two formulae is that the denominator for $t_{df=9}$ contains $S_X^2 = \frac{\Sigma(X - \bar{X})^2}{n-1}$ whereas the denominator of $Z_{\bar{X}}$ contains σ_X^2 (the known variance of the population from which the sample was drawn). This means that for a given sample of Xs, the value of S_X^2 computed on the sample might be either larger, smaller or exactly equal to σ_X^2 . While it may not be immediately obvious, the odds that S_X^2 will be smaller than σ_X^2 are much higher than the odds that it will be larger than σ_x^2. This is because the sampling distribution of s_x^2 is skewed and while, as we saw earlier, its mean is σ_x^2, its median is $< \sigma_x^2$

Here, for example, is the sampling distribution of s_x^2 for samples of size 10.

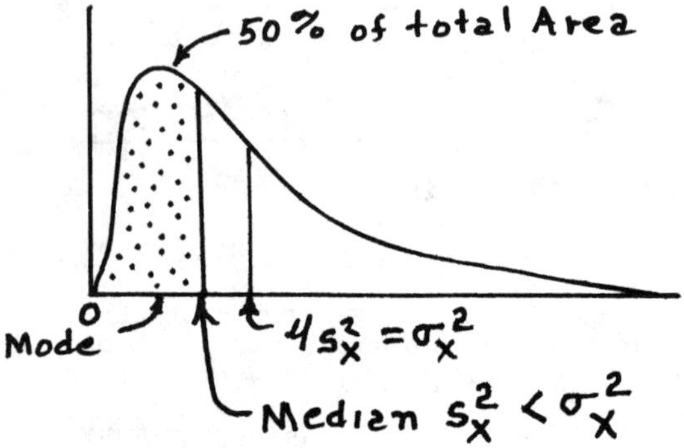

As can be seen, the median falls to the left of the mean. This is a general characteristic of the distributions that are skewed in the positive direction. Its importance here stems from the fact that for any distribution the median is the single value that divides the distribution into two equal areas. But if the median of the sampling distribution of s_x^2 is less than σ_x^2 it must follow that more than 50% of the sampling distribution of s_x^2 will have values that are less than σ_x^2. Another way to say the same thing is to assert that when we calculate s_x^2 on a given sample we are more likely to obtain a value of s_x^2 that is smaller than σ_x^2, than a value that is larger than σ_x^2. With some thought it should become apparent that this will happen even though μs_x^2 will exactly equal σ_x^2.

These considerations lead us to the conclusion that since s_x^2 appears in the demoninator of t, the value of t computed on a given sample will tend to have a larger absolute value than the comparable value of $z_{\bar{x}}$ that we would compute if σ_x^2 was known.

Another way to say this is to assert that the sampling distribution of t will be more variable than the comparable sampling distribution of $z_{\bar{X}}$. Moreover, since we know that $\sigma^2_{z_{\bar{X}}} = 1$ we can also assert that when df is small σ_t^2 must be greater than 1 and that σ_t^2 must approach 1 (as a limit) as the df for t increases.

Table V shows the critical values that cut off the upper .05, .025, .01 and .005 of the sampling distribution that is obtained when random samples of size n are drawn from a normal population where $\mu_X = \mu_h$

and $$t_{df=n-1} = \frac{\bar{X} - \mu_h}{\sqrt{\dfrac{s_X^2}{n}}}$$

is calculated on each sample.

To save space only the critical values for the upper end of the several t distributions have been tabled, but since we know that when $\mu_X = \mu_h$, the sampling distribution of t will be symmetrical about 0, we also know that the tabled values, preceded by a minus sign, are the critical values for the lower .05, .025, .01, and .005 of the sampling distribution of t.

By now it should be apparent to the thoughtful student that regardless of sample size, when we use

$$t_{df=n-1} = \frac{\bar{X} - \mu_h}{\sqrt{s_X^2/n}}$$

to test the statistical hypothesis that $\mu_X = \mu_h$ our risk of rejecting the hypothesis tested when it is true, will exactly equal the level we chose for α. From the material discussed so far it should also be apparent that our procedure in carrying out a test of the hypothesis that $\mu_X = \mu_h$ when σ_X^2 is unknown would have the same format as the $z_{\bar{X}}$ test that we use when σ_X^2 is known.

Here is an illustrative example:

Example IV - Testing an exact hypothesis about the mean, σ_x^2 unknown

Illustration

We wish to determine if Factory A is producing light bulbs of substandard quality (average burning time is greater than or less than some specific value); in this instance the variance of the factory output is unknown. Again we would test the hypothesis that the factory output is of standard quality, but this time we have to estimate the variance of the population of all light bulbs from the information in a sample.

Method

(1) Hypothesis: Population mean equals some specified value (μ_h). (Variance unknown)

(2) Set α

(3) Use $$t_{df=n-1} = \frac{\bar{X} - \mu_h}{\sqrt{\frac{s_x^2}{n}}}$$

where n = Sample size

$\bar{X}$ = Sample mean

μ_h = Hypothesized population mean

s_x^2 = Estimate of population variance

Note: $$s_x^2 = \frac{\sum(X - \bar{X})^2}{n-1}$$

(4)

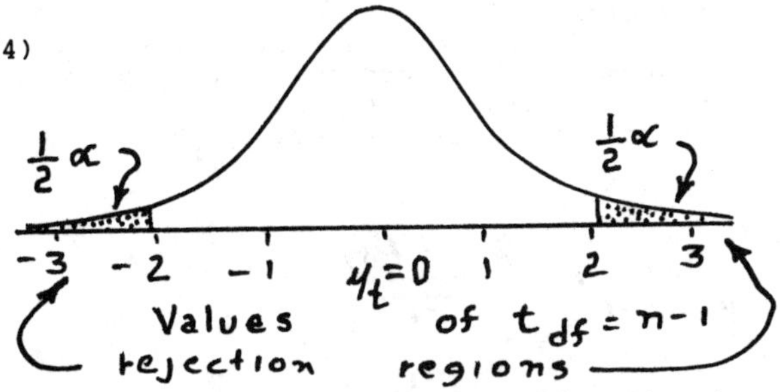

Determine the sampling distribution of t for df = n - 1, when the hypothesis tested is true.

Specify the rejection region or regions

(5) Compute t and determine whether or not it falls in a rejection region.

Notice that in the illustration, if we can reject the hypothesis that Factory A is producing standard quality bulbs, we will have reason to believe that Factory A is producing bulbs of substandard quality.

There is an additional point to be made about the statistic t that provides an important illustration of the logical coherence of the procedures we have been discussing. As we have seen, the formula that we use to test the hypothesis that $\mu_x = \mu_h$ when σ_x^2 is not known is:

$$t_{df=n-1} = \frac{\bar{X} - \mu_h}{\sqrt{\frac{s_x^2}{n}}}$$

We have just seen why, when our sample comes from a normal population and our statistical hypothesis is true (i.e., $\mu_x = \mu_h$), the sampling distribution of t has a mean of zero and we also saw why it begins to increasingly approximate a normal distribution with variance = 1, as sample size increases. It may not be immediately apparent, but it can also be shown that the sampling distribution of t is a special case of the sampling distribution of F. More specifically, it can be shown that

$$t^2_{df=n-1} = F_{df=1,\,n-1}$$

To see what this is true we need to first note that

$$t^2_{df=n-1} = \left[\frac{\bar{X} - \mu_h}{\sqrt{\frac{s_x^2}{n}}}\right]^2 = \frac{(\bar{X} - \mu_h)^2}{\frac{s_x^2}{n}}$$

In other words, t^2 consists of a ratio in which the numerator $= (\bar{X} - \mu_h)^2$ and the demoninator $= \frac{s_x^2}{n}$.

Earlier, we saw that F also consists of a ratio. Moreover, we saw that for F both the numerator and denominator of F are unbiased estimates of σ_x^2 when the two samples come from the same population. To see why $t^2 = F$ let's again begin with a normal population of Xs, with mean equal μ_X and variance equal σ_X^2

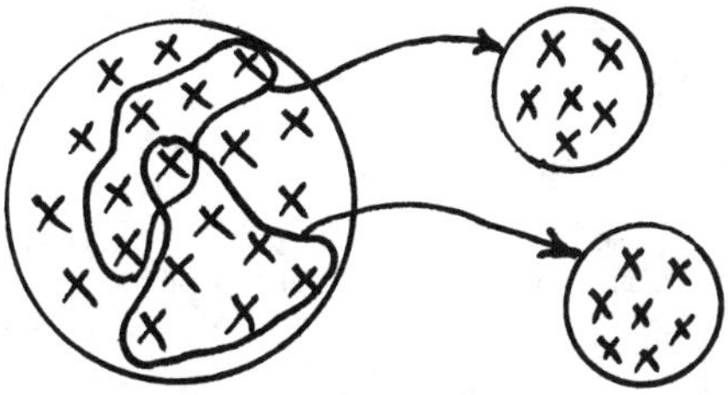

We saw earlier that if we drawn two random samples from this population and calculate s_X^2 on each sample, the statistic F is defined as follows:

$$F_{df=(n_1-1),(n_2-1)} = \frac{s_{X_1}^2}{s_{X_2}^2}$$

Suppose, however, that instead of calculating s_X^2 on the first sample, we calculate $\frac{\sum(X-\mu_X)^2}{n}$. We saw in Chapter 5 that like s_X^2, the statistic $\frac{\sum(X-\mu_X)^2}{n}$ is also an unbiased estimate of σ_X^2 and that where the df for s_X^2 is n - 1, the df for $\frac{\sum(X-\mu_X)^2}{n}$ is n.

This fact implies that we can also write the formula for F as follows:

$$F_{df=n_1,\, n_2-1} = \frac{\frac{\sum(X-\mu_X)^2}{n}}{s_{X_2}^2}$$

Let's use this alternative formula for F in the following situation. We draw two random samples from the above normal population of Xs with variance

$= \sigma_x^2$ and mean $= \mu_x$, but this time while the second sample consists of n randomly selected items, the first sample consists of only one randomly selected item.

We could not, of course, calculate $\dfrac{\sum (x-\bar{x})^2}{n-1}$ for a sample of 1 item because division by zero is an inadmissable arithmetic operation, but there is nothing to prevent us from calculating $\dfrac{\sum (x-\mu_x)^2}{n}$ on a sample where n = 1. When we do so we discover the obvious, namely that for n = 1

$$\frac{\sum (x-\mu_x)^2}{n} = (x-\mu_x)^2$$

In short, we discover that if $\dfrac{\sum (x-\mu_x)^2}{n} \rightarrow \sigma_x^2$ and if n = 1 the quantity $(x-\mu_x)^2$ is also an unbiased estimate of σ_x^2. All of this must mean that the statistic $\dfrac{(x-\mu_h)^2}{s_x^2}$ is distributed as $F_{df=1,n-1}$ when the items come from a normal population with $\mu_x = \mu_h$

It only requires one more conceptual step to see why the quantity

$$\frac{(\bar{x}-\mu_h)^2}{s_x^2/n}$$

must also be distributed as $F_{df=1,\ n-1}$. The important point to recognize is that the statistic

$\frac{(\bar{X} - \mu_h)^2}{s_X^2/n}$ consists of a ratio of two quantities $(\bar{X} - \mu_h)^2$ and $\frac{s_X^2}{n}$. With a little thought it should become apparent that if $\mu_X = \mu_h$ both quantities are unbiased estimates of σ_X^2. All of this implies that if we draw a single random sample of n items from a normal distribution where with mean = μ_h and we calculate the statistic

$$\frac{(\bar{X} - \mu_h)^2}{s_X^2/n}$$

we should be able to use the tabled values of $F_{df = 1, n-1}$ to determine if our obtained statistic is among those infrequent large values that occur either .05 (Table III) or .01 (Table IV) of the time. But if this is true it must also be true that the statistic

$$t^2_{df = n-1} = \frac{(\bar{X} - \mu_h)^2}{s_X^2/n} = F_{df = 1, n-1}$$

as we set out to show.

We can easily verify this conclusion for our selves by asking if the tabled values for $F_{df = 1, n - 1}$ are, in fact the square of the appropriate critical values in the corresponding distribution of $t_{df = n - 1}$

Let's try the value of t that cuts off the upper .025 of the sampling distribution of t. From Table V we find that the critical value for $t_{df = 9}$ is 2.262. Since, however, we also know that $t_{df = 9} = -2.262$ cuts off the lower .025 of the sampling distribution of t and since we would get the same value of t^2 if our obtained value of t was either -2.262 or +2.262, the probability of obtaining a value of $t^2 = (2.262)^2$

= 5.12 must be .025 + .025 = .05 when our sample comes from a normal population where $\mu_x = \mu_h$.

When we look into Table III we discover that the value of $F_{df = 1, 9}$ that cuts off the upper .05 of the F distribution is 5.12. Similarly, we can ascertain from Table V that since $t_{df = 21} = \pm 2.831$ cuts off the upper and lower .005 of the sampling distribution of t, an obtained value of $t^2_{df = 21} = (2.831)^2 = 8.02$ must be the critical value of $F_{df = 1, 21}$ that cuts off the upper .01 of the F distribution. When we look into Table IV we find that this is, in fact, the case.

Chapter 8

TESTS OF HYPOTHESES ABOUT THE MEANS OF TWO POPULATIONS WHEN $\sigma^2_{X_1}$ AND $\sigma^2_{X_2}$ ARE KNOWN

In Chapter 4 we saw how the statistic

$$Z_{\bar{X}} = \frac{\bar{X} - \mu_h}{\sqrt{\frac{\sigma_X^2}{n}}}$$

can be used to test an exact hypothesis about μ_X when

σ_X^2 is is known. We can use a comparable statistic:

$$Z_{(\bar{X}_1 - \bar{X}_2)} = \frac{(\bar{X}_1 - \bar{X}_2) - (\mu_{h_1} - \mu_{h_2})}{\sqrt{\frac{\sigma_{X_1}^2}{n_1} + \frac{\sigma_{X_2}^2}{n_2}}}$$

To test an exact hypothesis about the value of $(\mu_{X_1} - \mu_{X_2})$ (the difference between the means of two populations) when $\sigma^2_{X_1}$ and $\sigma^2_{X_2}$(variances of the two populations) are both known.

Our procedure would be quite comparable to the procedure we use when $Z_{\bar{X}}$ is employed to test an exact hypothesis about μ_X. When we use $Z_{(\bar{X}_1 - \bar{X}_2)}$, however we draw two random samples, one from each of the populations, and we compute $\bar{X} = \frac{\Sigma X}{n}$ on each. We then enter the obtained value of $\bar{X}_1 - \bar{X}_2$ into the formula for $Z_{(\bar{X}_1 - \bar{X}_2)}$ and we calculate the value of that results. If the hypothesis tested is true, that is if $\mu_{X_1} - \mu_{X_2}$ equals the value of $\mu_{h_1} - \mu_{h_2}$ that we use to calculate $Z_{(\bar{X}_1 - \bar{X}_2)}$ and if the samples come from normal populations (with known variances equal to $\sigma_{X_1}^2$ and $\sigma_{X_2}^2$, the sampling distribution of the statistic $Z_{(\bar{X}_1 - \bar{X}_2)}$ will be normal with $\sigma^2_{Z_{(\bar{X}_1 - \bar{X}_2)}} = 1$ and $\mu_{Z_{(\bar{X}_1 - \bar{X}_2)}} = 0$. This means that we can use Table I (the Z score table) to determine if our calculated value of $Z_{(\bar{X}_1 - \bar{X}_2)}$ is among those extreme values that only occur α proportion of the time when the hypothesis tested is true. If our calculated $Z_{(\bar{X}_1 - \bar{X}_2)}$ falls in a rejection region we reject the hypothesis tested at the α level of confidence. If the calculated $Z_{(\bar{X}_1 - \bar{X}_2)}$ does not fall in a rejection region we have no basis to reject the hypothesis tested.

Let us now proceed to carefully examine the formula for $Z_{(\bar{X}_1 - \bar{X}_2)}$ to see why all of this must be true.

Here is the formula for $Z_{(\bar{X}_1 - \bar{X}_2)}$

$$Z_{(\bar{X}_1 - \bar{X}_2)} = \frac{(\bar{X}_1 - \bar{X}_2) - (\mu_{h_1} - \mu_{h_2})}{\sqrt{\dfrac{\sigma_{X_1}^2}{n_1} + \dfrac{\sigma_{X_2}^2}{n_2}}}$$

The first point to note is that there is nothing about this formula that places a restriction on our sample sizes (i.e., n_1 and n_2 can be as large or as small as we wish).

Thus where $$Z_{\bar{X}} = \frac{\bar{X} - \mu_h}{\sqrt{\frac{\sigma_X^2}{n}}}$$

reduces to $$Z_X = \frac{X - \mu_h}{\sigma_X}$$ when $n = 1$

the formula for $Z_{(\bar{X}_1 - \bar{X}_2)}$

reduces to $$Z_{(X_1 - X_2)} = \frac{(X_1 - X_2) - (\mu_{h_1} - \mu_{h_2})}{\sqrt{\sigma_{X_1}^2 + \sigma_{X_2}^2}}$$

when $n_1 = n_2 = 1$

Let's now consider a situation where one might profitably use the statistic $Z_{(X_1 - X_2)}$. Imagine that we have two normal populations with known variances ($\sigma_{X_1}^2$ and $\sigma_{X_2}^2$) but with unknown means.

Here are those populations:

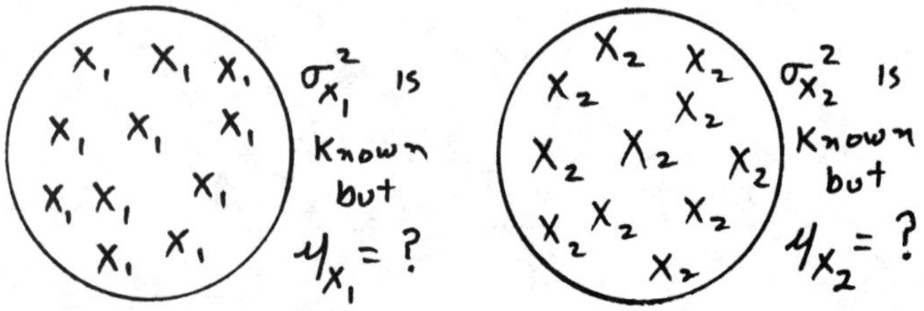

We can draw a single randomly selected item from each population and use the statistic

$$z_{(X_1 - X_2)} = \frac{(X_1 - X_2) - (\mu_{h_1} - \mu_{h_2})}{\sqrt{\sigma^2_{X_1} + \sigma^2_{X_2}}}$$

to test the statistical hypothesis that $(\mu_{X_1} - \mu_{X_2})$ equals any specific value $(\mu_{h_1} - \mu_{h_2})$ that we choose.

To do so, however, we have to figure out what the sampling distribution of $z_{(X_1 - X_2)}$ is when the hypothesis tested is true.

As a first step in that process let's ask what the sampling distribution of the statistic $X_1 - X_2$ is like when both X_1 and X_2 are randomly selected from a single normal popuation with mean = μ_X and variance = σ^2_X.

We know (from the material on pages 50-54 in Chapter 3) that when X_1 and X_2 come from the same normal population the sampling distribution of the statistic $X_1 + X_2$ is normal with

$$\mu_{(X_1 + X_2)} = \mu_{X_1} + \mu_{X_2}$$

and with

$$\sigma^2_{(X_1 + X_2)} = \sigma^2_{X_1} + \sigma^2_{X_2}$$

(At the point it may be helpful to review the material on pages 50-54 to remind ourselves exactly how we established these propositions).

Now let's consider what would happen if we again randomly select two items from the same normal population but this time we form the statistic

$$(X_1 + (-1)X_2) = (X_1 - X_2)$$

That is, we multiply the obtained value of X_2 by minus 1. From the rules for transforming Xs we can deduce that if the mean of the sampling distribution of X_2 is μ_{X_2} the mean of the sampling distribution of (-1) X_2 will equal $-\mu_{X_2}$. Those rules also tell us that if the variance of the sampling distribution of X_2 is $\sigma^2_{X_2}$ the variance of the sampling distribution of (-1) X_2 will still be $\sigma^2_{X_2}$. This is because the transformation rules tell us that if we multiply every item in a distribution by a constant we multiply the variance of the distribution by the square of the constant and in this instance the constant is (-1) and

(-1) = 1. All of this means that while

$$\mu(X_1 - X_2) = \mu X_1 - \mu X_2$$

$$\sigma^2_{(X_1 - X_2)} = \sigma^2_{X_1} + \sigma^2_{X_2}$$

In other words, contrary to what might have been our initial intuition, the variance of the sampling distribution of $X_1 - X_2$ is $\sigma^2_{X_1} + \sigma^2_{X_2}$ not $\sigma^2_{X_1} - \sigma^2_{X_2}$. Of course, if we think about it, we soon realize that $\sigma^2_{X_1} - \sigma^2_{X_2}$ couldn't possibly equal $\sigma^2_{(X_1 - X_2)}$ because if X_1 and X_2 both came from the same population it would imply that $\sigma^2_{(X_1 - X_2)}$ was equal to zero and for this to occur we would have to always get the same values of X_1 and X_2 when we randomly select our two items. Obviously, that is not going to happen if σ^2_X has any value >0.

Now let's consider the situation where instead of selecting X_1 & X_2 from a single population we randomly select X_1 from a normal population with mean equal to μX_1 and variance $= \sigma^2_{X_1}$ and we randomly select X_2 from a second normal population with mean $= \mu X_2$ and variance $= \sigma^2_{X_2}$. With some thought it should become clear that the principles of sampling that we have been studying ought to apply even when X_1 and X_2 come from populations with different means and different variances and that if the populations are normal to begin with, the sampling distribution of the statistic $X_1 - X_2$ will also be normal with

$$\mu_{(X_1 - X_2)} = \mu_{X_1} - \mu_{X_2}$$

and

$$\sigma^2_{(X_1 - X_2)} = \sigma^2_{X_1} + \sigma^2_{X_2}$$

Of course if $X_1 - X_2$ is distributed normally the statistic

$$Z_{(X_1 - X_2)} = \frac{(X_1 - X_2) - (\mu_{h_1} - \mu_{h_2})}{\sqrt{\sigma^2_{X_1} + \sigma^2_{X_2}}}$$

must also be distributed normally. Moreover its mean will be zero and its variance will be one.

To see why, we need only consider that when we compute $Z_{(X_1 - X_2)}$ on a given pair of items from populations where $(\mu_{X_1} - \mu_{X_2}) = (\mu_{h_1} - \mu_{h_2})$ we are converting the obtained value of $X_1 - X_2$ into a Z score where

An item in a sampling distribution of $(X_1 - X_2)$

The hypothesized mean of the sampling distribution of $(X_1 - X_2)$

$$Z_{(X_1 - X_2)} = \frac{(X_1 - X_2) - (\mu_{h_1} - \mu_{h_2})}{\sqrt{\sigma^2_{X_1} + \sigma^2_{X_2}}}$$

The standard deviation of the sampling distribution of $(X_1 - X_2)$

and we proved earlier (in Chapter 1) that when we convert every item in a distribution into Z scores we get a new distribution with a mean of zero and a variance of one.

It is all of these factors in combination that tell us that when our items come from normal populations in which

$$(\mu_{X_1} - \mu_{X_2}) = (\mu_{h_1} - \mu_{h_2})$$

is true, the table of Z scores (i.e., Table I) indicates the probabilities of attaining various values of Z by chance.

Of course we would seldom have the occasion to want to test the hypothesis that

$$(\mu_{X_1} - \mu_{X_2}) = (\mu_{h_1} - \mu_{h_2})$$

by using only one item from each population. Ordinarily, when we wanted to test this hypothesis we would draw random samples of n_1 and n_2 items from the populations in question and compute

$$z_{(\bar{X}_1-\bar{X}_2)} = \frac{(\bar{X}_1 - \bar{X}_2) - (\mu_{h_1} - \mu_{h_2})}{\sqrt{\frac{\sigma_{X_1}^2}{n_1} + \frac{\sigma_{X_2}^2}{n_2}}}$$

but the logic of our test would be exactly the same as when $n_1 = n_2 = 1$. If our samples come from normal populations with $(\mu_{X_1} - \mu_{X_2}) = (\mu_{h_1} - \mu_{h_2})$, the statistic $z_{(\bar{X}_1 - \bar{X}_2)}$ would be exactly analogous to the statistic $z_{(X_1 - X_2)}$. However, in this instance we would have:

An item in the sampling distribution of $(\bar{X}_1 - \bar{X}_2)$

The hypothesized mean of the sampling distribution of $(\bar{X}_1 - \bar{X}_2)$

$$z_{(\bar{X}_1-\bar{X}_2)} = \frac{(\bar{X}_1 - \bar{X}_2) - (\mu_{h_1} - \mu_{h_2})}{\sqrt{\frac{\sigma_{X_1}^2}{n_1} + \frac{\sigma_{X_2}^2}{n_2}}}$$

The standard deviation of the sampling distribution of $(\bar{X}_1 - \bar{X}_2)$

In short, as suggested earlier, when our samples come from normal populations where the hypothesis that

$(\mu_{x_1} - \mu_{x_2}) = (\mu_{h_1} - \mu_{h_2})$ is true, we can use

the table of Z scores to determine the critical values in our statistical test.

Here is an illustrative example.

Example V - Testing an Hypothesis about two means (Variances known)

Illustration

We wish to decide if teaching Method 1 produces different results from teaching Method 2. In order to do so we test the statistical hypothesis that the two methods produce the same results. (i.e., our two samples came from populations with the same means.) Variances are assumed to be known.

Method

(1) Hypothesis: Population means equal

i.e., $(\mu_{x_1} - \mu_{x_2}) = (\mu_{h_1} - \mu_{h_2}) = 0$

(2) Set α

(3) Use

$$Z_{(\bar{X}_1 - \bar{X}_2)} = \frac{(\bar{X}_1 - \bar{X}_2) - (\mu_{h_1} - \mu_{h_2})}{\sqrt{\frac{\sigma^2_{x_1}}{n_1} + \frac{\sigma^2_{x_2}}{n_2}}}$$

$$= \frac{\bar{X}_1 - \bar{X}_2}{\sqrt{\frac{\sigma^2_{x_1}}{n_1} + \frac{\sigma^2_{x_2}}{n_2}}}$$

μ_{h_1} = Hypothesized population mean Method 1

μ_{h_2} = Hypothesized population mean Method 2

$\bar{X}_1$ = Sample mean Method 1

$\bar{X}_2$ = Sample mean Method 2

$\sigma^2_{X_1}$ = Known population variance Method 1

$\sigma^2_{X_2}$ = Known population variance Method 2

n_1 = Sample size Method 1

n_2 = Sample size Method 2

(4)

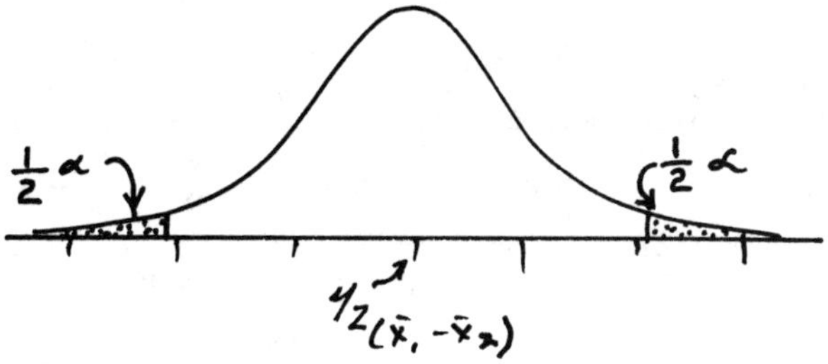

Sampling distribution of $z_{(\bar{x}_1 - \bar{x}_2)}$

under the assumption that the hyothesis tested is true.

(5) Compute $z_{(\bar{x}_1 - \bar{x}_2)}$ and determine whether or not it falls in a rejection region.

Notice in the illustration that if we can reject the hypothesis that there is no difference between the two methods, we will have reason to believe that there is a difference.

Chapter 9

TESTS OF HYPOTHESES ABOUT THE MEANS OF TWO POPULATIONS WHEN $\sigma^2_{X_1}$ AND $\sigma^2_{X_2}$ ARE NOT KNOWN

We have just seen how we can use the statistic

$$z_{(\bar{X}_1 - \bar{X}_2)} = \frac{(\bar{X}_1 - \bar{X}_2) - (\mu_1 - \mu_2)}{\sqrt{\frac{\sigma^2_{X_1}}{n_1} + \frac{\sigma^2_{X_2}}{n_2}}}$$

computed on two random samples to test a statistical hypothesis about the means of the populations from which the samples were drawn. To carry out that test, however, we had to know the values of $\sigma^2_{X_1}$ and $\sigma^2_{X_2}$ (the variances of the two populations). When we do not have this information at our disposal we can use the s^2_Xs computed on our samples to estimate the population variances and use the statistic t to carry out our statistical test.

$$t_{(\bar{X}_1 - \bar{X}_2)} = \frac{(\bar{X}_1 - \bar{X}_2) - (\mu_1 - \mu_2)}{\sqrt{\frac{s^2_{X_1}}{n_1} + \frac{s^2_{X_2}}{n_2}}}$$

At the outset it is important to recognize that

while the statistic $z_{(\bar{X}_1 - \bar{X}_2)}$ can (theoretically at least) be used in situations in which $\sigma^2_{X_1} \neq \sigma^2_{X_2}$ several statisticians have suggested that use of the statistic $t_{(\bar{X}_1 - \bar{X}_2)}$ be restricted to situations in which there is no reason to suspect that $s^2_{X_1}$ and $s^2_{X_2}$ are not estimating the same quantity. In other words, it has been suggested that $t_{(\bar{X}_1 - \bar{X}_2)}$ only be used in situations in which one believes that the two samples were drawn from populations with the same variance.

The problem arises from the fact that when $\sigma^2_{X_1}$ does not equal $\sigma^2_{X_2}$ the denominator of $t_{(\bar{X}_1 - \bar{X}_2)}$ contains two estimates of two quantities and this raises the question as to just how these two estimates are to be properly combined. Fortunately we do not have to face this problem when $s^2_{X_1}$ and $s^2_{X_2}$ are both estimating the same quantity. When this happens we can combine the information in our two samples to derive a single estimate of σ^2_X with df $= (n_1 - 1) + (n_2 - 1)$ To do so, we calculate what is described as a weighted average of the s^2_X s that are computed on our two samples.

One way to write the formula for this kind of weighted average is

$$\frac{(n_1 - 1)s^2_{X_1} + (n_2 - 1)s^2_{X_2}}{(n_1 - 1) + (n_2 - 1)}$$

Another expression which says the same thing is

$$s_p^2 = \frac{\sum(X_1-\bar{X}_1)^2 + \sum(X_2-\bar{X}_2)^2}{(n_1-1)+(n_2-1)}$$

In essence, when we calculate s_p^2 (called a pooled estimate of the variance) we are combining the sums of squared deviations from the means and dividing by the combined degrees of freedom. When $n_1 = n_2$

$$s_p^2 = \frac{s_{X_1}^2 + s_{X_2}^2}{2}.$$

i.e., s_p^2 is merely the average of the two s_X^2s. When $n_1 \neq n_2$, however, s_p^2 will not ordinarily equal the average of the two s_X^2s. Instead, as indicated above, it will represent a weighted average where the sample with the larger df will make a larger contribution to the estimate than the sample with the smaller df.

Earlier it was noted that the df for s_p^2 was $(n_1-1)+(n_2-1)$. This should be apparent from several consideratons. For example: when we combine $s_{X_1}^2$ and $s_{X_2}^2$ to form an estimate of σ_X^2 we are combining two statistics that have (n_1-1) and (n_2-1) df to form a single estimate of a single quantity. It seems only reasonable that our combined estimate should have the combined df of the two estimates considered individually. Moreover, if we examine the formula for s_p^2 we discover that while it contains $n_1 + n_2$ values of $(X-\bar{X})$, only n_1+n_2-2 of them are free to vary. In short, we lose a degree of freedom for each of the two estimates of μ_X (i.e., the $\bar{X}$s) that appeared in the formula for s_p^2 .

All of this implies that if we have reason to believe that $s_{X_1}^2$ and $s_{X_2}^2$ are both estimates of the same value of σ_X^2, the df for a t test of the hypothesis that $\mu_{X_1} = \mu_{X_2}$ would be $n_1 + n_2 - 2$ and the formula for t could be written as:

$$t = \frac{(\bar{X}_1 - \bar{X}_2) - (\mu_{n_1} - \mu_{n_2})}{\sqrt{s_p^2\left(\frac{1}{n_1} + \frac{1}{n_2}\right)}}$$

Moreover, if we wished to test the hypothesis that $\mu_{X_1} - \mu_{X_2} = 0$ the formula for t would take the following form:

$$t = \frac{\bar{X}_1 - \bar{X}_2}{\sqrt{s_p^2\left(\frac{n_1 + n_2}{n_1 n_2}\right)}}$$

The next page shows two other ways to write this same expression. Each is the algebraic equivalent of the above formula and each appears in one textbook or another.

Note: $$\frac{n_1 + n_2}{n_1 n_2} = \frac{1}{n_1}\left(\frac{n_2}{n_2}\right) + \frac{1}{n_2}\left(\frac{n_1}{n_1}\right)$$
$$= \frac{1}{n_1} + \frac{1}{n_2}$$

$$t_{(df = n_1 + n_2 - 2)}$$

$$= \frac{\dfrac{\sum X_1}{n_1} - \dfrac{\sum X_2}{n_2}}{\sqrt{\dfrac{\Sigma X_1 - \dfrac{(\Sigma X_1)^2}{n_1} + \Sigma X_2 - \dfrac{(\Sigma X_2)^2}{n_2}}{n_1 - n_2 - 2}\left(\dfrac{1}{n_1} + \dfrac{1}{n_2}\right)}}$$

$$t_{(df = (n_1 - 1) + (n_2 - 1))}$$

$$= \frac{\bar{X}_1 - \bar{X}_2}{\sqrt{\dfrac{\Sigma X_1 - \dfrac{(\Sigma X_1)^2}{n_1} + \Sigma X_2 - \dfrac{(\Sigma X_2)^2}{n_2}}{n_1 + n_2 - 2}\left(\dfrac{n_1 + n_2}{n_1 n_2}\right)}}$$

The derivation of the denominator of these formula make use of the fact that for any set of n items:

$$\sum (x - \bar{X})^2 = \sum x^2 - \frac{(\sum x)^2}{n}$$

This can be shown as follows:

$$\sum (x - \bar{X})^2$$

$$= \sum (x^2 - 2\bar{X}X + \bar{X})^2$$

Just as $(a - b)^2 = (a^2 - 2ab + b^2)$

$$= \sum x^2 - 2\bar{X}\sum X + n\bar{X}^2$$

Note that for a given set of Xs $\bar{X}^2$ is a constant, hence we must add it n times

$$= \sum x^2 - 2\frac{\sum X}{n}\sum X + n\frac{(\sum X)^2}{n^2}$$

$$= \sum x^2 - 2\frac{(\sum X)^2}{n} + \frac{(\sum X)^2}{n}$$

$$= \sum x^2 - \frac{(\sum X)^2}{n}$$

Earlier it was noted that the statistic $t_{(\bar{X}_1 - \bar{X}_2)}$ has its major application in situations where we believe our samples come from normal populations in which $\sigma^2_{X_1} = \sigma^2_{X_2}$. If the obtained $s^2_{X_1}$ and $s^2_{X_2}$ are reasonably close to each other, and if the distributions in the two samples are approximately normal it is likely that these assumptions will have been met. If, however, $s^2_{X_1}$ and $s^2_{X_2}$ are quite different. For example, if a two-tailed F test yields a significant value, or if either of our samples departs appreciably from normal, it is likely that we will have failed to meet at least one of the assumptions for the use of the t statistic. This raises the question of how the sampling distribution of t is changed when the assumptions for its use are violated.

One way to answer this question is to program a computer to draw a pair of random samples from populations with characteristics that violate the assumptions for t in known ways and to compute an obtained value of t on the data. By doing so repeatedly, the computer produces an empirical approximation to the sampling distribution of t from the populations being studied.

When this was done it was found that even when $\sigma^2_{X_1} \neq \sigma^2_{X_2}$ and even when the populations were not normal, under certain conditions the empirical sampling distribution of $t = \dfrac{\bar{X}_1 - \bar{X}_2}{\sqrt{s^2_p\left(\frac{1}{n_1} + \frac{1}{n_2}\right)}}$ was very similar to the theoretical sampling distribution for t with df = $n_1 + n_2 - 2$. For this to occur, however, it was important that $n_1 = n_2$ and that the samples contain at least 25 items.

Apparently, if we use reasonably large samples and take care to use equal sample sizes, we can pool our S_X^2 s, even though we know that $\sigma_{X_1}^2$ is not in fact equal to $\sigma_{X_2}^2$ and the samples come from populations that are not normal.

We can gain some insight into why we can do so if we consider that when $n_1 = n_2$ the formula

$$s_p^2 \left[\frac{1}{n_1} + \frac{1}{n_2} \right]$$

is numerically equal to

$$\left[\frac{s_{X_1}^2}{n_1} + \frac{s_{X_2}^2}{n_2} \right]$$

In other words, it will make no difference if we pool or not. The obtained value of t will be the same when $n_1 = n_2$

We also know from the central limit theorem that even when our original population is not normal, if n is large, the sampling distibution of $\bar{X}$ will tend to approxmate a normal distribution. This helps to explain why the computer simulations revealed that with large n's the t statistic was relatively insensitive to violation of the assumption that the populations be normally distributed.

These considerations point to the conclusion that when using t we have much to gain if we can arrange to use equal n's and if we can keep them reasonably large. In this respect, it is important to note that in addition to the advantages cited above, there is an arithmetic advantage to be gained from using equal n's. The important factor here is that when, for practical purposes, the number of items in our two samples must be restricted to some specific total, the quantity $\frac{1}{n_1} + \frac{1}{n_2}$ will be minimized when we arrange that $n_1 = n_2$. For example, if we must randomly assign each of 70 subjects to one of two groups a t test on the difference between the means of the groups will be most sensitive when

$n_1 = n_2 = 35$. Thus $\frac{1}{35} + \frac{1}{35} = .057$ is smaller than $\frac{1}{10} + \frac{1}{60} = .117$

Finally, we can gain even more insight into the t test for a difference between means if we again take note of the relationship between t and F.

We saw earlier that

$$t^2_{df=(n-1)} = F_{df\ 1,\ (n-1)}$$

when $\mu_X = \mu_h$

A similar equivalence occurs for t computed on the difference between means when the samples are drawn from the same population.

Thus

$$t^2_{df=(n_1+n_2-2)} = F_{df=1,(n_1+n_2-2)}$$

when $\mu_{X_1} - \mu_{X_2} = \mu_{h_1} - \mu_{h_2} = 0$

To see why, it will be helpful to first note that when we square the above formula we find that

$$t^2 = \frac{[(\bar{X}_1 - \bar{X}_2) - 0]^2}{\left(\frac{s_p^2}{n_1} + \frac{s_p^2}{n_2}\right)}$$

From the material covered earlier, we should by now be able to recognize that the denominator of this expression $\left(\frac{s_p^2}{n_1} + \frac{s_p^2}{n_2}\right)$ is an unbiased estimated of $\sigma^2_{(\bar{X}_1 - \bar{X}_2)}$ with $df = n_1 + n_2 - 2$. The numerator is also an unbiased estimate of $\sigma^2_{(\bar{X}_1 - \bar{X}_2)}$ but with df = 1. This is because the obtained $[(\bar{X}_1 - \bar{X}_2) - 0]^2$ represents one item in the sampling distribution of $[(\bar{X}_1 - \bar{X}_2) - 0]^2$. Since we know that when the samples come from the same population, $\mu[(\bar{X}_1 - \bar{X}_2) - 0]^2 = \sigma^2_{(\bar{X}_1 - \bar{X}_2)}$ we know that a single $[(\bar{X}_1 - \bar{X}_2) - 0]^2 \rightarrow \sigma^2_{(\bar{X}_1 - \bar{X}_2)}$.

All of this means that t^2 computed on the difference between the means of two random samples from a given normal population consists of a ratio of two unbiased estimates of $\sigma^2_{(\bar{X}_1 - \bar{X}_2)}$.

In other words,

$$t^2_{df = (n_1 + n_2 - 2)} = F_{df = 1, (n_1 + n_2 - 2)}$$

This leads us to Illustrative Example VI.

Example VI - Test of a hypothesis about the means of two populations, $\sigma^2_{X_1}$ and $\sigma^2_{X_2}$ not known

Illustration

Each of a number of rats is assigned randomly to one of two groups and tested in a learning task. Rats in both groups receive 20 trials, but for Group 1 the trials occur at 30 second intervals, whereas for Group 2 the interval between trials is 5 minutes. We find that the average of the performances of the rats in Group 2 is superior to the average of the performances of the rats in Group 1. We must now decide whether or not the observed difference between $\bar{X}_1$ and $\bar{X}_2$ can be attributed to the difference in the intertrial intervals for the rats in the two groups or whether, alternatively, the observed difference can be attributed to the kind of variations that inevitably occur when we take random samples.

Method

(1) Hypothesis: $\mu_{x_1} - \mu_{x_2} = \mu_{n_1} - \mu_{n_2} = 0$

(2) Set α

(3) Use $t_{df=(n_1+n_2-2)} = \dfrac{\bar{X}_1 - \bar{X}_2}{\sqrt{s^2_p\left(\frac{1}{n_1}+\frac{1}{n_2}\right)}}$

where

n_1 = Size of Sample 1

n_2 = Size of Sample 2

$\bar{X}_1$ = Mean of Group 1

$\bar{X}_2$ = Mean of Group 2

$$s_p^2 = \frac{\sum(X_1 - \bar{X}_1)^2 + \sum(X_2 - \bar{X}_2)^2}{n_1 + n_2 - 2}$$

(4)

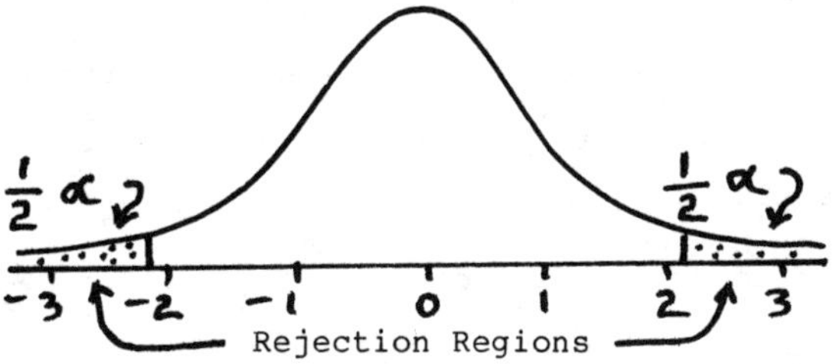

Sampling distribution of t under the assumption that the hypothesis tested is true.

(5) Compute the obtained value of t and determine if it falls in one of the rejection regions. If it does, we will have reason to believe that the observed difference between $\bar{X}_1$ and $\bar{X}_2$ is due to the difference in intertrial intervals for the two groups.

Notice this test is predicated on the assumption that both samples come from the same population and hence that $\sigma^2_{X_1} = \sigma^2_{X_2}$. If an F test on $s^2_{X_1}$ and $s^2_{X_2}$ led us to reject the hypothesis that $\sigma^2_{X_1} = \sigma^2_{X_2}$, we could still use the above procedure, if our sample sizes were equal and were reasonably large (i.e., n_s = 25 or more).

Chapter 10

THE CONCEPT OF β AND THE POWER OF A TEST

In previous chapters we have seen that when testing a given statistical hypothesis we always run a certain risk that the data in our sample (or samples) will lead us to reject that hypothesis when, in fact, that hypothesis is true. Statisticians described such an event as a Type I error. With the procedures we have been discussing, the probability of a Type I error is exactly equal to the value selected for α.

Whenever we conduct a statistical test there is also a possibility that our data will lead us to commit a second kind of error. A Type II error would occur if we failed to reject the hypothesis tested when in fact that hypothesis was false. Statisticians use the symbol β to represent the probability of a Type II error. Unlike α, which is set by the investigator the value of β is determined by a number of interlocking factors, one of which is the value selected for α. An example can help to explain why this is so.

Earlier (on pages 67-68) we explored the use of $z_{\bar{X}}$ to test the statistical hypothesis that μ_X (the average burning time of all light bulbs produced in a given factory) was equal to a specific value (740 hrs).To test this hypothesis we took a random sample of 25 bulbs and assessed their burning times. In the example (on page 68) it was found that the mean for the sample was 728 hours.

We then entered this value into the formula

$$z_{\bar{X}} = \frac{\bar{X} - \mu_h}{\sqrt{\frac{\sigma_X^2}{n}}}$$

where n = sample size (i.e., 25)

$\bar{X}$ = the mean of the sample (i.e., 728 hrs)

μ_h = the hypothetical value of μ_x (i.e., 740 hrs)

σ_x^2 = the known variance of burning times for all bulbs (i.e., 400 hrs)

When we did so we found that

$$z_{\bar{x}} = \frac{728 - 740}{\sqrt{\frac{400}{25}}} = -3$$

From the table of Z scores we found that for a two-tailed test with α = .05 the critical values for $z_{\bar{x}}$ were $\pm$ 1.96. Since the obtained $z_{\bar{x}}$ was -3.0 it fell into the lower rejection region. This led us to reject the statistical hypothesis that the population from which the sample was drawn had a mean = 740 hours.

It will be helpful to carefully re-examine these ideas when they are depicted graphically.

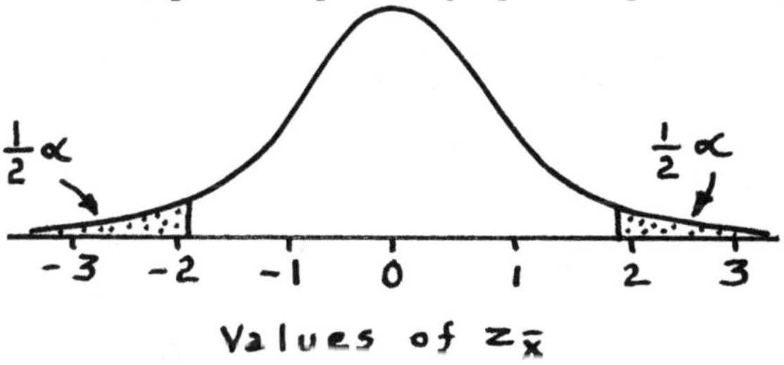

This is the theoretically derived sampling distribution for $z_{\bar{x}}$ based on the assumption that we take random samples of size n = 25 from a normal population with known variance = 400 and mean = 740.

Since the obtained value of $z_{\bar{x}}$ = -3 is among those extreme values that would only occur 5% of the time when the statistical hypothesis is true, we reject that hypothesis and conclude that the equipment is producing bulbs with a mean burning time that is some value other than 740 hours. That is, we conclude that the equipment is producing substandard bulbs.

We don't, of course, know the actual value of the population mean, but we at least can have some confidence that it's not 740. How much confidence is indicated by the level of α we elected to use in our test. Since we set α = .05 we can be confident that if the population mean had been 740, something happened in our test that would only occur 5% of the time that we took samples of size n = 25.

What, however, could we have concluded had our test yielded a less extreme value of $Z_{\bar{X}}$; for example a value that was larger than - 1.96 but less than + 1.96? (Such a value would be among those that would occur 95% of the time when the hypothesis tested was true.) Under those circumstances we would have had no basis to reject the hypothesis tested. But what then should we have done? We might have been tempted to accept that hypothesis, but as will be seen, unless we had previously specified a particular alternative hypothesis, we would have had no way of calculating the risk entailed in doing so. We can see why if we choose an alternative hypothesis and proceed to calculate that risk.

The next page shows two sampling distributions of the statistic

$$Z_{\bar{X}} = \frac{\bar{X} - \mu_h}{\sqrt{\frac{\sigma_x^2}{n}}} = \frac{\bar{X} - 740}{\sqrt{\frac{400}{25}}}$$

when samples are drawn from a population in which the hypothesis tested is true and when it is false by a given amount.

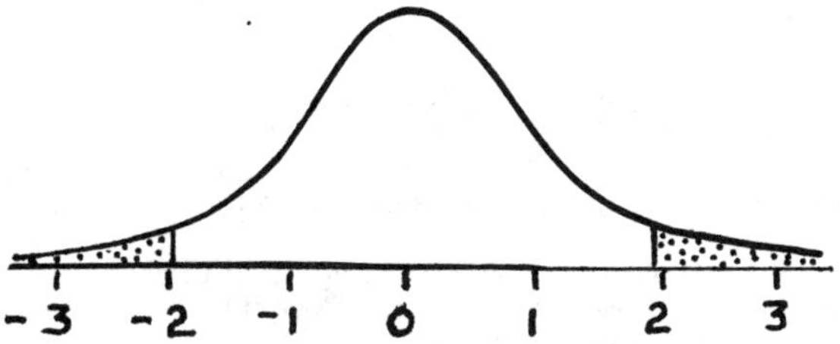

Values of $Z_{\bar{X}}$

This is the sampling distribution of $Z_{\bar{X}}$ when samples of size n = 25 are drawn from a normal population with $\sigma_X^2 = 400$ and with $\mu_X = 740$. That is, this is the sampling distribution of $Z_{\bar{X}}$ when the hypothesis tested is true and $\mu_X = 740$.

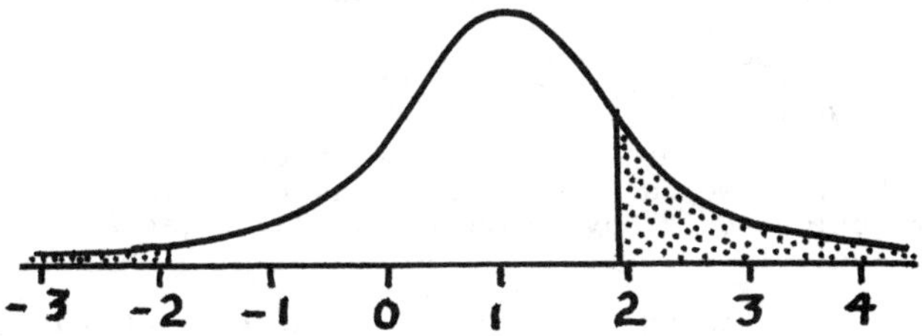

Values of $Z_{\bar{X}}$

This is the sampling distribution of $Z_{\bar{X}}$ when samples of size n = 25 are drawn from a normal population with $\sigma_X^2 = 400$ but with $\mu_X = 744$. That is, this is the sampling distribution of $Z_{\bar{X}}$ when the hypothesis tested ($\mu_X = 740$) is false by a given amount.

Notice that the mean of the second distribution is + 1 rather than zero. We can see how this comes about if we consider the sampling distribution of $\bar{X}$ that would result if samples of size n = 25 were drawn from a normal population with $\sigma_X^2 = 400$ and $\mu_X = 744$.

From the Central Limit Theorem we can deduce that the resulting sampling distribution would be normal with

$$\mu_{\bar{X}} = \mu_X = 744$$

$$\sigma_{\bar{X}}^2 = \frac{\sigma_X^2}{n} = \frac{400}{25} = 16$$

Now let's consider how this sampling distribution would be transformed if we subtracted 740 from each $\bar{X}$. From the rules for transforming distributions (discussed earlier) we can deduce that the resulting distribution would have a mean equal to 4 and a variance = 16.

> (Subtracting a constant from every item in a distribution subtracts the constant from the mean but it leaves the variance unchanged.)

Next, let us consider how the distribution of the statistic ($\bar{X}$ - 740) would be transformed if every item in that distribution was divided by

$$\sqrt{\frac{400}{25}} = \sqrt{16} = 4$$

Again from the rules for transforming distributions we can deduce that the resulting distribution would have a mean = 1 and a standard deviation = 1.

> (Dividing every item in a distribution by a constant divides the mean by that constant and it divides the variance by the square of the constant and it divides the standard deviation by the constant.)

In summary, we have now deduced that if we draw samples of size n = 25 from a normal population with σ_X^2 = 400 and μ_X = 744 and then calculate the statistic

$$z_{\bar{X}} = \frac{\bar{X} - 740}{\sqrt{\frac{400}{25}}}$$

on each sample we will obtain a sampling distribution that is normal with a standard deviation of 1 and a mean of 1.

This is the distribution that was depicted on the bottom of Page 155. It is the distribution of values of

$$Z_{\bar{X}} = \frac{\bar{X} - 740}{\sqrt{\frac{400}{25}}}$$

that would obtain if we were testing the statistical hypothesis that $\mu_X = 740$ when in fact that hypothesis is false and the alternative hypothesis that $\mu_X = 744$ is true.

The next page again shows the sampling distribution of

$$Z_{\bar{X}} = \frac{\bar{X} - 740}{\sqrt{\frac{400}{25}}}$$

when the hypothesis tested (i.e., $\mu_X = 740$) is true as well as when the alternative hypothesis ($\mu_X = 744$) is true.) This time, however, the areas of each of the critical regions of these distributions have been specified exactly.

Examination of these areas reveals that the probability of a Type II error (accepting the hypothesis tested when the alternative hypothesis is true) is exactly .83.

It will be informative to check this calculation for oneself. Note that in the lower distribution on the next page an item with a value of + 1.96 is in fact only .96 σ above the mean of its own distribution, whereas an item with a value of - 1.96 is in fact 2.96 σ below that mean. By using the table of Z scores one can determine that the area in a normal distribution bounded by ordinates at - 2.96 and + .96 σ is .83 of the total area.

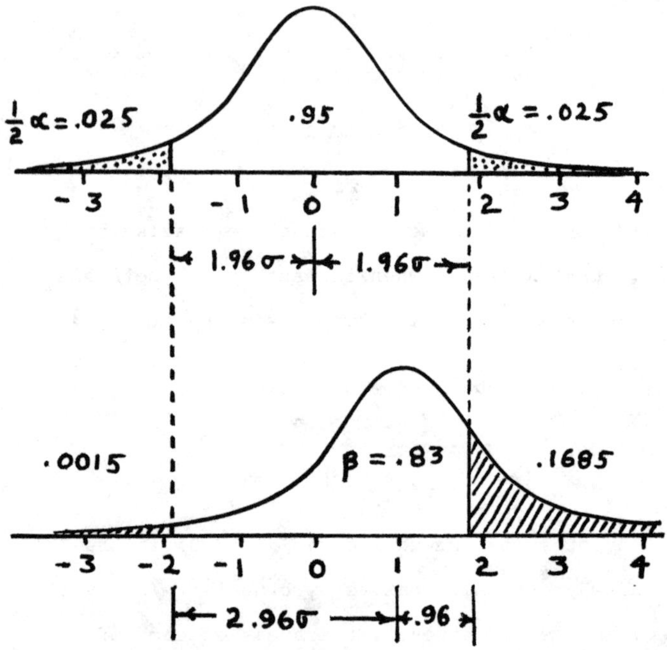

As seen in the above Figure, β represents the probability that the statistical procedure will lead to accepting the hypothesis tested when a specific alternative hypothesis is true. Some reflection on this fact should make it clear that the quantity $1 - \beta$ must represent the probability of rejecting the hypothesis tested when the alternative hypothesis is true. This probability is called Power and like β, Power is determined by a variety of factors in addition to the level at which α has been set.

The next page again shows the sampling distribution of $Z_{\bar{X}}$ when the hypothesis tested (i.e., $\mu_X = 740$) is true as well as when it is false. This time, however, the alternative hypothesis is that $\mu_X = 748$, rather than $\mu_X = 744$.

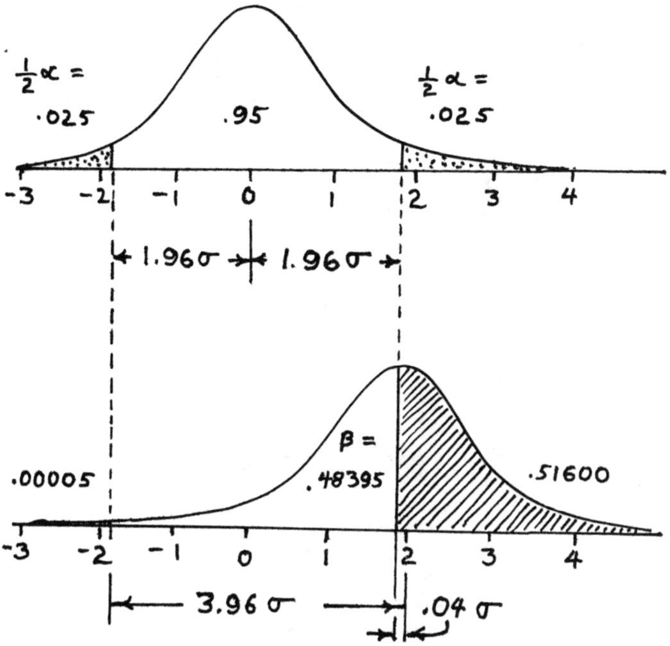

Notice that in the lower distribution an item with a value of + 1.96 is in fact .04 σ below the mean whereas an item with a value of - 1.96 is in fact 3.96 σ below the same mean. This is because (1.96 x 2) + (2 - 1.96) = 3.96. Using the table of Z scores we find that the area bound by ordinates at - 3.96 and - .04 σ from the mean is .49995 - .0160 = .48395. Accordingly we can conclude that when testing the hypothesis that μ_X = 740 against the alternative hypothesis that μ_X = 748, the chance of a Type II error (i.e., β) is .48395 and the power of the test is 1 - .48395 = .51605.

We can summarize these as well as our earlier observations by asserting that with the statistic used here (i.e., $z_{\bar{x}}$) and with a sample of size n = 25 and with α = .05, β will equal .83 if the hypothesis tested (i.e., μ_x = 740) is false and the alternative hypothesis that μ_x = 744 is true. On the other hand, if the more extreme alternative hypothesis that μ_x = 748 is true β will equal .48 rather than .83.

These calculations are represented graphically by the starred items in the next figure.

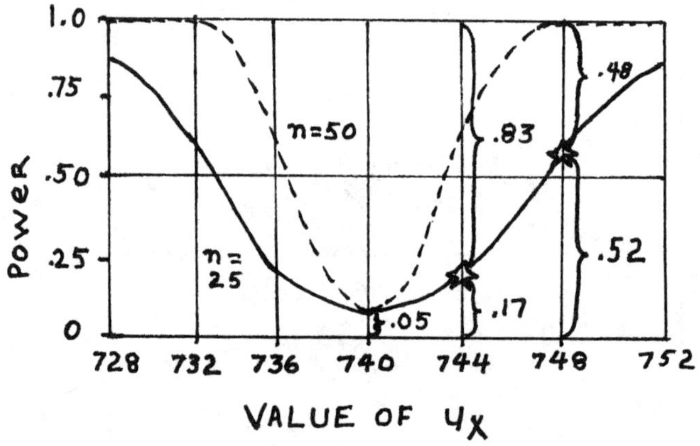

Notice how this figure is arranged. The ordinate shows the probability that the sample will reject the hypothesis (i.e., the power of the test) that μ_x = 740 when μ_x has the value indicated on the abscissa.

Notice that when μ_x equals the hypothesis tested, the value for power is exactly equal to α (.e., when μ_x = 740, power = .05)

Furthermore, since power = 1 - β , the power of the test will be (1 - .83) =. 17 when μ_x = 744 and it will be (1 - .48) = .52 when μ_x = 748.

Of course these values (.17 and .52) are but two of the infinite set of values of power that can be calculated for this particular statistical test. The continuous line (called a power function) shows many of these values when one draws samples of size n = 25.

The dotted line shows the power function for the same test when samples of size n = 50 rather than n = 25 are drawn. Notice how the slope of the power function increases with an increase in sample size.

As can be seen, the larger the sample size the more sensitive is the test to departures from the hypothesis being tested. Thus, when n = 25 and when μ_x is 748, power is only .52, but power is nearly 1.0 when μ_x = 748 and n = 50.

This direct relationship between sensitivity and sample size has an important implication for how one should interpret a given statistical test that appears in the scientific literature. It suggests that if a very large sample size has been employed in conducting that test, one should be on guard for the possibility that the test has picked up the effect of an extraneous (i.e., confounding) variable, rather than the variable or effect under investigation. Conversely, however, if sample size is relatively small, a statistically significant result would suggest that the effect under investigation is probably relatively large and hence that one's results are unlikely to reflect the action of some unrecognized confounding factor.

Chapter 11

CORRELATION AND REGRESSION

Before engaging in a discussion of correlation and regression it is necessary to point out that any student who takes the time to look into various textbooks on statistics will quickly realize that there is no uniformity with respect to the use of symbols. In general, this reflects the fact that many of the concepts in statistics are relatively new and, in statistics as in many other areas of human endeavor, it takes a great deal of time before uniform practices emerge. Unfortunately, this set of circumstances greatly complicates the student's task for it means that one simply cannot take a given author's symbols at face value, and must, in all cases, make certain that he or she understands exactly what the author means by the particular symbols used.

Up to now we have gone to considerable trouble to make it clear that in this book the symbol μ_X refers to the mean of a set of items when those items are treated as a population. In this context, we have also asserted that the symbol $\bar{X}$ is used to represent the mean of a set of items when the set is treated as a sample from a given population. We have already seen why this convention is helpful when we must make inferences about a population based on the information in a sample. Unfortunately, however, many of the textbooks that deal with the topics of correlation and regression fail to keep this convention in mind and they use the symbol $\bar{X}$ to represent the mean of a set of items even though they are treating the data they are analyzing as if it were a population.

In order to maximize the logical coherence of the present text we will avoid this practice. More specifically, we will approach the topic of correlation and regression as a problem in descriptive statistics in which we are seeking reasonable ways to summarize and quantify the relationships that may exist within a given set of data (i.e., within a given population). Thus, if the data are a set of Xs we will use the symbols $\mu_X = \frac{\sum X}{N}$ and $\sigma_X^2 = \frac{\sum (X - \mu_X)^2}{N}$ to signify the mean and variance of those Xs.

At the outset it is important to emphasize that the concepts of correlation and regression are only applicable to situations in which we have a set of N identifiable items, each of which yields two measures (X and Y). For example, the items might be people and the X measures might be a given person's height while the Y measure might be that person's weight. In these circumstances, a given item (John Jones, for instance) might turn out to be 66 inches tall and weigh 154 pounds. Another person (Jack Smith), on the other hand, might be 65 inches tall and weigh 152 pounds.

Here is a set of N = 6 items along with the two measures on each of them.

	Item Number	X	Y
	1	63	151
John Jones	2	66	154
	3	63	152
	4	67	154
Jack Smith	5	65	152
	6	64	154

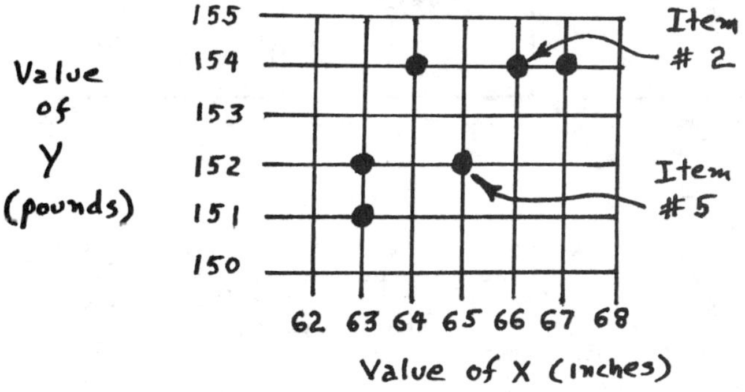

As seen on the previous page, we can graphically represent these data by plotting them in such a way that each point represents both the X and the Y measure for a given item.

With a large number of items we may expect the graph (called a scatterplot) to take one of many forms. Here are some examples.

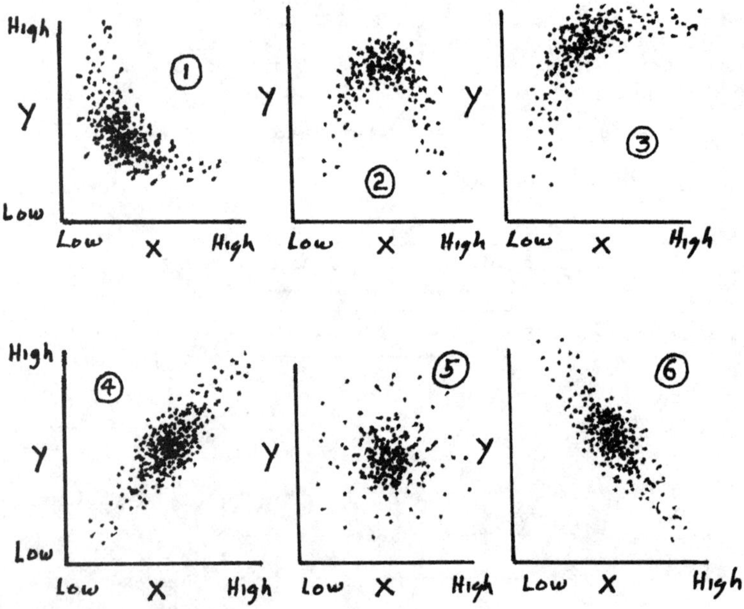

In this book we shall deal only with problems of correlation and regression for data that yield points which tend to fall along a straight line (as in plots 4 and 6). We will not deal with the problems that arise when the points tend to fall along a curved line (as in 1, 2, & 3). Put another way, we shall only deal with data in which the relationship between X and Y is linear. Of course there are methods for dealing with nonlinear relationships but the topic is beyond the scope of this book.

If we examine plots 4 and 6 we see that the values of X and Y tend to covary in a systematic fashion. Thus, in plot 4, items with high X values tend also to have high Y values. In plot 6, items with high X values tend to have low Y values. Finally, it is noteworthy that in plot 5 there appears to be no consistent relationship between the X and Y values for the several items.

Quantifying the Relationship I
(The coefficient of correlation)

Here are two scatterplots of the height vs. weight of a large number of subjects.

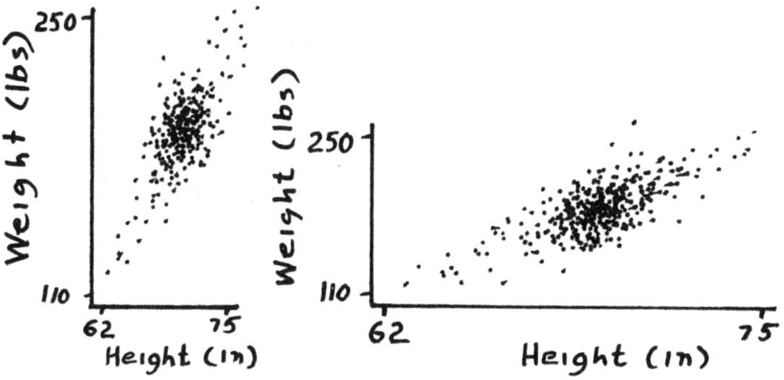

Both scatterplots are based on the same set of measures and in both, items with high X values tend to also have high Y values. But for the left hand scatterplot the Y values appear to be much more variable than the X values, whereas in the right hand scatterplot the X values appear to be more variable than the Y values. We know, however, that despite these visual differences the two scatterplots must be exhibiting an identical relationship between X and Y and that these impressions are largely a product of the fact that the X and Y measures are not in the same units and this means that we can arbitrarily assign as much or as little scale as we please to each measure.

One way to make an equitable assignment of these scales is to plot our data in units that are the standard deviations of our two sets of measures. To do so, we merely convert each of the X and Y values into Z scores.

Thus Z_X for a given item = $\dfrac{X - \mu_X}{\sigma_X}$

where $\mu_X = \dfrac{\Sigma X}{N}$ $\qquad \sigma_X = \sqrt{\dfrac{\Sigma(X - \mu_X)^2}{N}}$

and Z_Y for that <u>same item</u> = $\dfrac{Y - \mu_Y}{\sigma_Y}$

where $\mu_Y = \dfrac{\Sigma Y}{N}$ $\qquad \sigma_Y = \sqrt{\dfrac{\Sigma(Y - \mu_Y)^2}{N}}$

Note that we used μ_X and μ_Y in the formula for Z_X and Z_Y. This is appropriate because we are treating our set of items as if it were a population. As noted earlier, we are approaching the problem of correlation as one in which we are basically trying to summarize and to quantify certain characteristics of a given set of data.

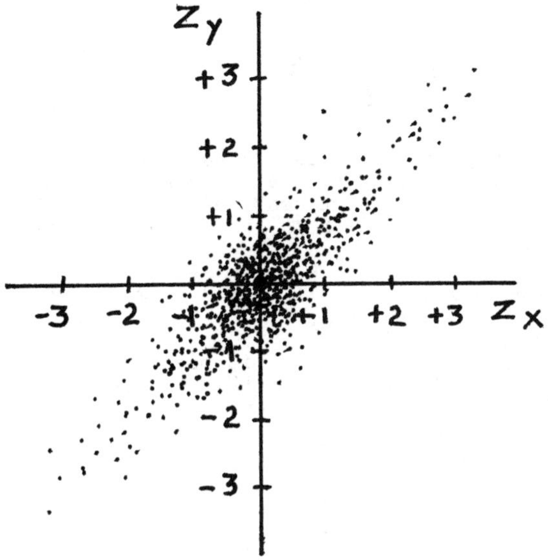

When we convert the previous measures of height and weight into Z scores we obtain the above scatterplot. As can be seen, when we arrange to plot our data in Z score form, we arrange that the transformed scores (Z scores) will have a mean of zero and a variance equal to one.

In addition to providing an intuitively reasonable way to visually represent correlational data, the Z score transformation enables us to readily obtain a quantitative index of the relationship between X and Y. That index is called (the correlation coefficient) and it is defined as the mean Z score product.

Thus, by definition $$r = \frac{\sum z_x z_y}{N}$$

where $$z_x = \frac{X - \mu_x}{\sigma_x} \qquad z_y = \frac{Y - \mu_y}{\sigma_y}$$

To see why the numerical value of r can serve as a quantitative index of the relationship between X and Y it will be helpful to carefully examine the coordinate system in which z_x and z_y are plotted.

This is Quadrant IV. For item in Quadrants IV z_x is negative and z_y is positive, therefore, the product ($z_x z_y$) is negative

This is Quadrant I. For Items in Quadrant I z_x is positive and z_y is positive, therefore, the product ($z_x z_y$) is positive.

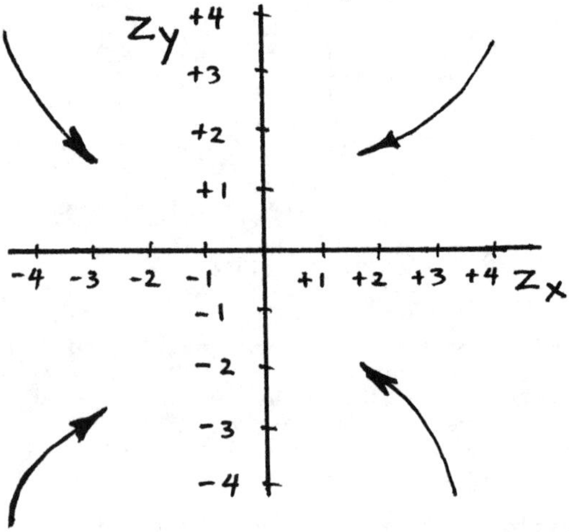

This is Quadrant III. For items in Quadrant III z_x and z_y are both negative, hence, the product ($z_x z_y$) is positive.

This is Quadrant II. For items in Quadrant II z_x is positive, but z_y is negative, therefore, the product ($z_x z_y$) is negative

Now let's look at three of the possible configurations that various sets of data might exhibit when displayed in Z score form.

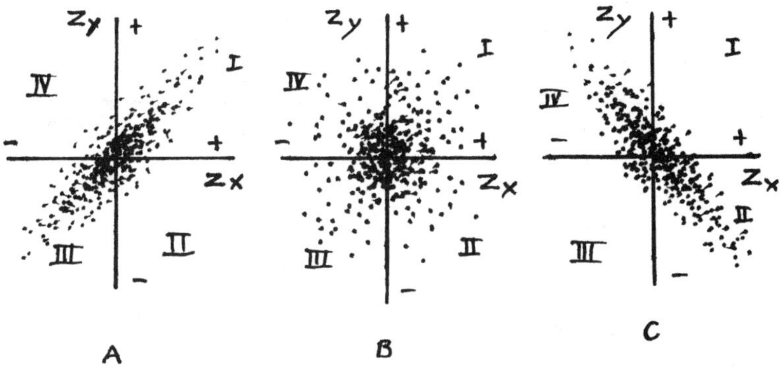

As can be seen, in Scatterplot A, most of the items fall in Quadrants I and III, hence their Z score products are positive and the value of r will be positive. In Scatterplot C most of the items fall in Quadrants II and IV, hence their Z score products are negative and accordingly, the value of r will be negative. In Scatterplot B the items tend to be equally distributed among the four quadrants, hence there are approximately equal numbers of positive and negative Z score products and their average $\left(\frac{\Sigma z_x z_y}{N}\right)$ will be near zero.

When X and Y have a perfect relationship all of the points in a scatterplot of the data will fall along a straight line.

Perfect positive relationship

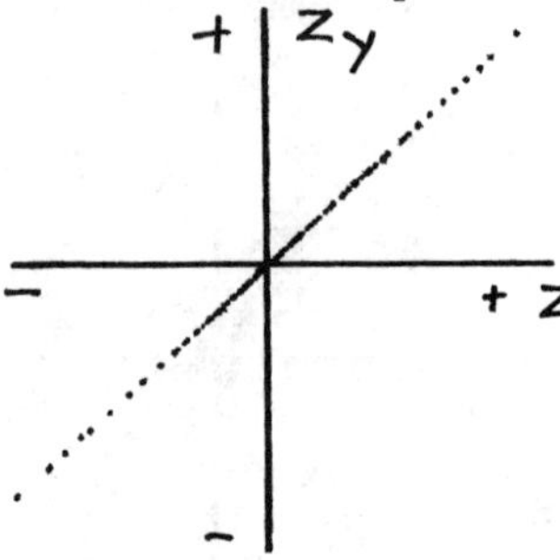

Perfect negative relationship

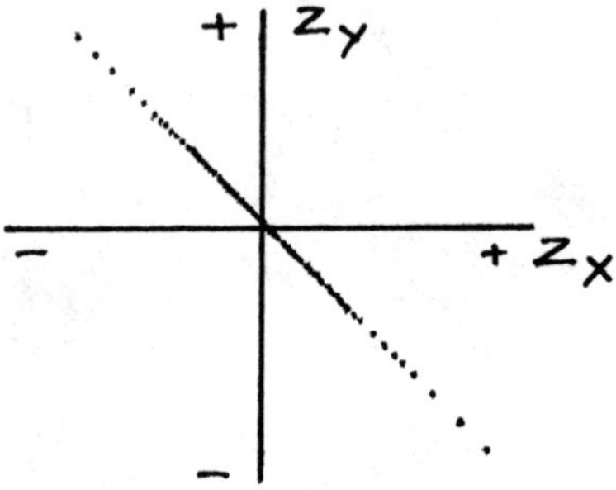

If the points fall along, a line each of the Z score products $(z_x z_y)$ will have the same sign; moreover, for each item the absolute value of z_x will equal the absolute value of z_y.

Consider a perfect positive (or negative) relationship.

Under these conditions, for each item $|z_x| = |z_y|$

and $|r| = \dfrac{\sum z_x^2}{N} = \dfrac{\sum z_y^2}{N}$

This observation is of special interest because, as we will see, for any distribution

$$\sum z^2 = N$$

$$\sum z^2 = \sum\left(\frac{X - \mu_x}{\sigma_x}\right)^2 = \frac{1}{\sigma_x^2}\sum(X-\mu_x)^2$$

$$\text{but } \frac{1}{\sigma_x^2} = \frac{1}{\frac{\sum(X-\mu_x)^2}{N}} = \frac{N}{\sum(X-\mu_x)^2}$$

$$\text{thus } \sum z^2 = \frac{N}{\sum(X-\mu_x)^2}\sum(X-\mu_x)^2 = N$$

But if in a perfect relationship $|r| = \frac{\sum z^2}{N}$ and if $\sum z^2 = N$ we can conclude that when the relationship between X and Y is perfect $|r|$ = N/N = 1.

All this means that the values of $r = \frac{\sum z_x z_y}{N}$ can only vary from -1 through 0 to +1.

When r = -1	When r = 0	When r = +1
we have a perfect negative relationship between X and Y	we have no relationship between X and Y	we have a perfect positive relationship between X and Y

There are times when it is convenient to express r in a form that is designed to make its numerical calculations more convenient. The next page shows the derivation of one of the various computational formula for r.

$$r = \frac{\Sigma z_x z_y}{N} = \frac{1}{N}\sum \frac{(X-\mu_x)(Y-\mu_y)}{\sigma_x \quad \sigma_x}$$

but $\sum (X-\mu_x)(Y-\mu_y)$

$$= \sum (XY - Y\mu_x - X\mu_y + \mu_x \mu_y)$$

$$= \Sigma XY - \mu_x \Sigma Y - \mu_y \Sigma X + N\left(\frac{\Sigma X}{N}\right)\left(\frac{\Sigma Y}{N}\right)$$

$$= \Sigma XY - \frac{\Sigma X \Sigma Y}{N} - \frac{\Sigma Y \Sigma X}{N} + \frac{(\Sigma X)(\Sigma Y)}{N}$$

$$= \Sigma XY - \frac{(\Sigma X)(\Sigma Y)}{N}$$

thus $r = \frac{1}{N\sigma_x \sigma_y} \Sigma (X-\mu_x)(Y-\mu_y)$

$$= \frac{\Sigma XY - \frac{(\Sigma X)(\Sigma Y)}{N}}{N \sigma_x \sigma_y}$$

but $N \sigma_x \sigma_y = \sqrt{N^2 \sigma_x^2 \sigma_y^2}$

and $\sigma_x^2 = \frac{\Sigma X^2}{N} - \frac{(\Sigma X)^2}{N^2}$

and $\sigma_y^2 = \frac{\Sigma Y^2}{N} - \frac{(\Sigma Y)^2}{N^2}$

$$r = \frac{\sum z_x z_y}{N}$$

$$= \frac{\sum xy - \frac{(\sum x)(\sum y)}{N}}{N \sigma_x \sigma_y}$$

$$= \frac{\sum xy - \frac{(\sum x)(\sum y)}{N}}{\sqrt{N^2\left(\frac{\sum x^2}{N} - \frac{(\sum x)^2}{N^2}\right)\left(\frac{\sum y^2}{N} - \frac{(\sum y)^2}{N^2}\right)}}$$

$$= \frac{\sum xy - \frac{(\sum x)(\sum y)}{N}}{\sqrt{\left(\sum x^2 - \frac{(\sum x)^2}{N}\right)\left(\sum y^2 - \frac{(\sum y)^2}{N}\right)}}$$

Quantifying the Relationship II
Correlation and the Concept of Regression

Correlation and regression are two aspects of a single phenomenon. They are both expressions of the relationship between X and Y when we have two measures on each of a set of N items.

Here is a set of items that can be used to illustrate the basic concepts of regression.

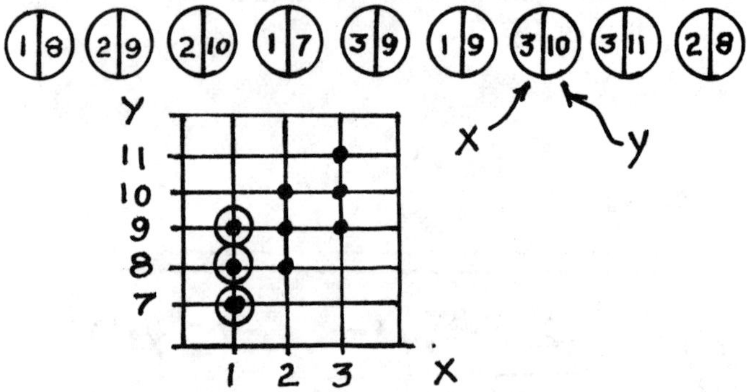

In this highly schematic illustration, N equals nine.

Notice that some items have the same X values as other items.

For instance, the items in which X is equal to 1 are circled. The Y values for these items are 7, 8 and 9 respectively.

The next figure shows the same data. This time, however, there is a square around the means of the Y values for each value of X. The figure also includes a line through these means. This line is known as a regression line.

To the left of the scatterplot is a frequency distribution of the Y values. This is the frequency distribution that would be obtained if for each item we had measured Y but not X. (Don't be disturbed by the fact that the distribution is on its side, it is plotted in this fashion merely because it is convenient to do so.)

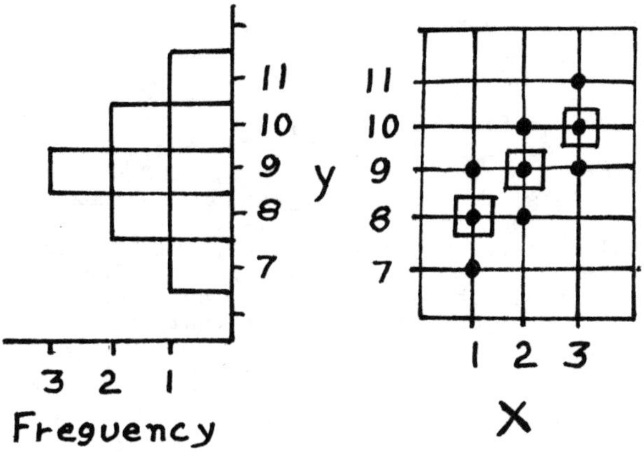

We saw in Chapter 3 that for any distribution, the sum of the squared deviations around the mean of the distribution is smaller than the sum of the squared deviations around any other point.

Accordingly, if we are about to draw an item at random from a given distribution, our best estimate of the value that the item will have is the mean of the distribution.

In the above example, if we draw an item at random and do not look at its X value, our best estimate of the Y value will be 9 (the mean of all the Y values). On the other hand, suppose that when we draw our item at random we are able to take note of its X value. If we have the scatterplot before us and we find that the X value for the item is 3, then our best estimate of the Y value would be 10 (the mean of all the Y values with X equal to 3).

If Y is the actual score of the item and μ_y is our best estimate when we do not look at the X value for the item, then the quantity Y - μ_y would be the error in prediction and the quantity

$$\sigma_Y^2 = \frac{\sum (Y - \mu_Y)^2}{N}$$

would be an intuitively reasonable index of the average error in predicting Y when we have no knowledge of X.

If, however, we had the scatterplot before us and we looked at the X score for the item, then our prediction of the item's Y score would be the mean of the Y values for those items with the same X value. If $\tilde{Y}$ equals that mean (i.e., $\tilde{Y}$ is the appropriate value on the regression line), and if Y is the actual score of the particular item in question, the difference between the actual Y value and $\tilde{Y}$ would be an error in prediction. Under these circumstances the quantity

$$\sigma_{err}^2 = \frac{\sum (Y - \tilde{Y})^2}{N}$$

would be an intuitively reasonable index of the average error in predicting Y from X across all items.

In the numerical example on page 174 there are nine items and

$$\mu_Y = 9$$

Accordingly,

$$\sigma_Y^2 = \frac{1}{9}\Big[(7-9)^2 + (8-9)^2 + (9-9)^2 + (8-9)^2 + (9-9)^2 + (10-9)^2 + (9-9)^2 + (10-9)^2 + (11-9)^2\Big]$$

$$= \frac{1}{9}\,[12] = 1.33$$

In the numerical example on page 174

when $X = 1$, $\tilde{y} = 8$

when $X = 2$, $\tilde{y} = 9$

when $X = 3$, $\tilde{y} = 10$

Accordingly $$\sigma^2_{err} = \frac{\sum (y - \tilde{y})^2}{N}$$

$$= \frac{1}{9}\Big[(7-8)^2 + (8-8)^2 + (9-8)^2 + (8-9)^2 + (9-9)^2 + (10-9)^2 + (9-10)^2 + (10-10)^2 + (11-10)^2\Big]$$

$$= \frac{1}{9}[6] = .66$$

Obviously, in this example, we run less chance of a large error if we take account of the X value of the item when we attempt to predict the item's Y value. If the X and Y scores for the various items were not covarying (in the example, high scores on X are associated with high scores on Y), we would obtain no such improvement in prediction.

This leads us to a second way to specify the correlation coefficient (r). We saw previously that by definition

$$r = \frac{\sum z_x z_y}{N}$$

We can (and later will) also show that for a given set of items the numerical value of r is equal to

$$\sqrt{1 - \frac{\sigma^2_{err}}{\sigma^2_y}}$$

Accordingly

$$r^2 = 1 - \frac{\sigma^2_{err}}{\sigma^2_y}$$

We have just found that in the numerical example based on the data on page 174

$$\sigma^2_{err} = .66 \quad \text{whereas} \quad \sigma_y^2 = 1.33$$

From these facts we can conclude that

$$r^2 = 1 - \frac{.66}{1.33} = .504$$

If as asserted above

$$r = \sqrt{1 - \frac{\sigma^2_{err}}{\sigma^2_y}}$$

Then for the numerical example of page 174

$$r = \sqrt{.504} = .71$$

We can check for ourselves whether or not this is the same value as is given by the formula

$$r = \frac{\sum z_x z_y}{N}$$

The next page shows the calculations of $r = \frac{\sum z_x z_y}{N}$ for the data on page 174. As can be seen, both approaches to r yield the same value.

z_x	$(X-\mu_x)$	X	Y	$(Y-\mu_y)$	z_y	$z_x z_y$
-1.23	-1	1	7	-2	-1.74	2.14
-1.23	-1	1	8	-1	-.87	1.07
-1.23	-1	1	9	0	0	0
0	0	2	8	-1	-.87	0
0	0	2	9	0	0	0
0	0	2	10	1	.87	0
1.23	1	3	9	0	0	0
1.23	1	3	10	1	.87	1.07
1.23	1	3	11	2	1.74	2.14
$\Sigma z_x = 0$	$\Sigma(X-\mu_x) = 0$	$\Sigma X = 18$	$\Sigma Y = 81$	$\Sigma(Y-\mu_y) = 0$	$\Sigma z_y = 0$	$\Sigma z_x z_y = 6.42$

$$\sigma_x^2 = \frac{1}{9}\left[(-1)^2 + (-1)^2 + (-1)^2 + (0)^2 + (0)^2 + (0)^2 + (1)^2 + (1)^2 + (1)^2\right]$$

$$= \frac{1}{9}[6] = .66$$

$$\sigma_x = \sqrt{.66} = .81$$

$$\sigma_y^2 = \frac{1}{9}\left[(-2)^2 + (-1)^2 + (0)^2 + (-1)^2 + (0)^2 + (1)^2 + (0)^2 + (1)^2 + (2)^2\right]$$

$$= \frac{1}{9}[12] = 1.33$$

$$\sigma_y = \sqrt{1.33} = 1.15$$

$$r = \frac{\Sigma z_x z_y}{N} = \frac{6.42}{9}$$

$$\boxed{= .71}$$

Specifying the regression line

In order to specify a given regression line we must identify the one line that minimizes the errors in estimating Y given X and to do so we use a least squares criterion. In other words, we desire that line for which the sum of the squared deviations of all the Y values is at a minimum.

While this line may often tend to pass near the means of the several Y values for each value of X, it would be surprising if for a given set of data the means of the Y values at each value of X fell exactly on the one line that satisfies the least squares criterion.

We saw (in Chapter 1) that the equation of a straight line has the form:

$$Y = bX + a$$

where a is the value of Y when X = 0 and b is the change in Y for a unit change in X.

We also saw that by substituting various values of X into this equation we can determine the value of Y for any given value of X.

Graphically we have:

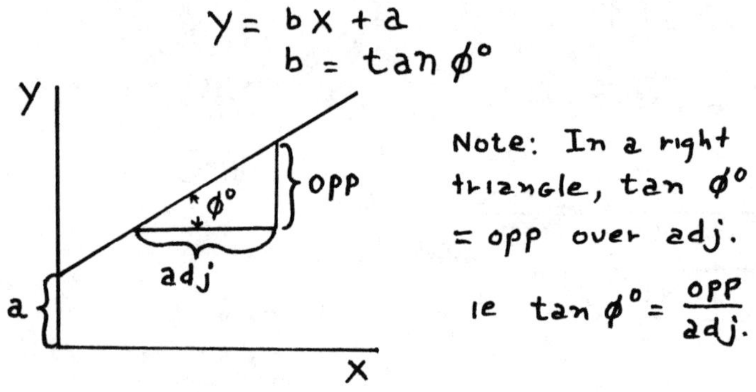

We form a regression equation by writing the expression $\tilde{Y} = bx + a$ where $\tilde{Y}$ is the predicted value of Y for a given value of X. If we are to write the regression equation for a given set of data, however, we must find the values of a and b which will define the one line that will minimize the errors in estimating Y given X. To do so, we enter our obtained values of N, ΣX, ΣY, etc., into the following set of equations and solve them for a and b.

$$a\Sigma X + b\Sigma X^2 = \Sigma XY$$

$$aN + b\Sigma X = \Sigma Y$$

The derivation of these equations is presented below for the student who is familiar with calculus. It is included in this book on the assumption that even if a student has no knowledge of calculus, it may be helpful to at least have a glimpse of the mathematical reasoning behind these equations.

As noted above, the regression line has the form $\tilde{Y} = bx + a$. For a given set of data each value of Y (for each value of X) will either equal $\tilde{Y}$ or deviate above or below $\tilde{Y}$. We will use the symbol d to represent this deviation.

Thus

$$y + d = \tilde{y}$$

$$y + d = a + bx$$

$$d = a + bx - y$$

$$d^2 = (a + bx - y)^2$$

When we sum across all N items in our data we find that

$$\Sigma d^2 = \Sigma (a + bx - y)^2$$

We must find the values of a + b that will minimize Σd^2. We can find these values if we let f

represent $\sum d^2$ and set the partial derivatives with respect to a and b equal to zero.

That is, we set

$$\frac{\partial f}{\partial a} = 2\sum(a + bX - Y) = 0$$

$$\frac{\partial f}{\partial b} = 2\sum(a + bX - Y)X = 0$$

If we divide by 2 and then carry out the summation we find that $\sum d^2$ is minimized when

$$\sum a + b\sum X - \sum Y = 0$$

$$a\sum X + b\sum X^2 - \sum XY = 0$$

Moreover, since a is a constant and we are summing across N items we have

$$Na + b\sum X - \sum Y = 0$$

and

$$a\sum X + b\sum X^2 - \sum XY = 0$$

or as seen above

$$a\sum X + b\sum X^2 = \sum XY$$

and $$Na + b\sum X = \sum Y$$

These are simultaneous equations and in high school algebra we learned how to solve them. We can refresh ourselves on this procedure if we examine their application to the data on page 174.

For those data:

$$N = 9 \qquad \Sigma X^2 = 42 \qquad \Sigma Y = 81$$

$$\Sigma X = 18 \qquad \Sigma Y^2 = 741 \qquad \Sigma XY = 168$$

When we enter these numbers into the above equations we find that

$$a(18) + b(42) = 168$$

and $a(9) + b(18) = 81$

In high school algebra we learned that we do not destroy an algebraic equality if we multiply the quantity on both sides of the equal sign by the same number.

Thus: $a(18) + b(42) = 168$

and $2(a(9) + b(18)) = 2(81)$

This means that

$$a(18) + b(42) = 168$$

and $a(18) + b(36) = 162$

We also do no harm to an algebraic equality if we subtract the same amount from both sides of the equal sign. Since $a(18) + b(36) = 162$ we leave the equality in the equation $a(18) + b(42) = 168$ intact if we subtract $a(18) + b(36)$ from the left side of the equation as long as we also subtract 162 from the right side of this equation.

Here are those operations:

$$\begin{array}{r} a(18) + b(42) = 168 \\ -a(18) - b(36) = -162 \\ \hline 0 \;\; + b(\;6) = + \;\;\; 6 \end{array}$$

This leads us to conclude that $b = 1$ but if $b = 1$ and if $a(18) + (1)42 = 168$ we can also conclude that

$$a = \frac{168 - 42}{18} = 7$$

All of this means that for the data on page 174 the values of a and b in the expression $\tilde{Y} = bx + a$ that will minimize the sum of squared deviations from the regression line are a = 7 and b = 1.

In other words, for the data on page 174 we can use the equation Y = X + 7 to determine the best estimates of Y given X. For example:

when X = 1 $\quad \tilde{Y} = 1 + 7 = 8$

when X = 2 $\quad \tilde{Y} = 2 + 7 = 9$

and when X = 3 $\quad \tilde{Y} = 3 + 7 = 10$

As can be seen, these are the same values of $\tilde{Y}$ that we obtained earlier when we calculated the means of the Y values for each value of X. This is as it should be for the data on page 174 because as seen on page 174, the regression line passes directly through the means of the Y values for each value of X. More often than not, however, the regression line calculated with the above equations will not pass directly through all of the means of the Y values for each value of X. Instead some will be above and some below the line that best describes their trend. Since, however, the above equations give us the values of a and b for the one line that will minimize the sum of squared deviations from itself, the line specified by those equations is the one we seek when we want to minimize our errors in estimating Y given X and that line will tend to pass through the means of the Y values for each value of X.

We can gain additional insight into correlation and regression if we examine the application of the above equations to the determination of the regression line when our Xs and Ys have been transformed into z_x s and z_y s.

Under such circumstances the equations

$$a\Sigma x + b\Sigma x^2 = \Sigma xy$$

and

$$aN + b\Sigma x = \Sigma y$$

become

$$a\sum z_x + b\sum z_x^2 = \sum z_x z_y$$

and

$$aN + b\sum z_x = \sum z_y$$

but we showed earlier that

$$\sum z_x = 0 = \sum z_y \quad \text{and} \quad \sum z_x^2 = N$$

Thus our equations reduce to

$$a(0) + bN = \sum z_x z_y$$

and

$$aN + b(0) = 0$$

i.e.,

$$bN = \sum z_x z_y$$

This means that

$$b = \frac{\sum z_x z_y}{N}$$

and

$$a = 0$$

All of this implies that if we plot our data in Z score form the regression line will pass through the intersection of $Z_{\bar{x}}$ and $Z_{\bar{y}}$ (which is zero), and its slope will be equal to r. Both of these facts are illustrated on the next page.

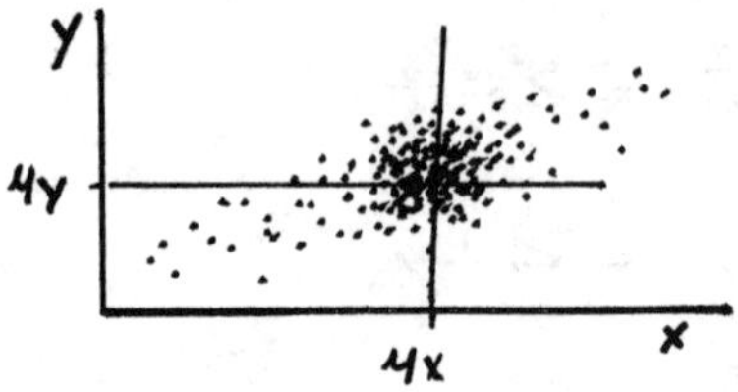

Here is a scatter-
plot of the X and
Y values for a
large number of
items

$z_{\tilde{y}} = (\tan \theta^{o}) z_x$

$\tan \theta^{o} = \dfrac{\sum z_x z_y}{N}$

$a = 0$

$z_{\tilde{y}} = r z_x$

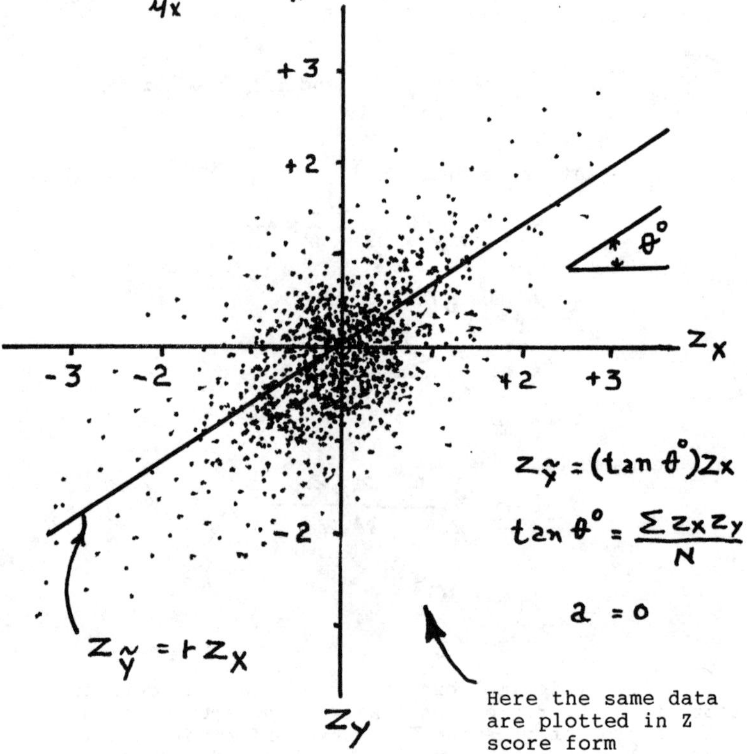

Here the same data
are plotted in Z
score form

Here are three scatterplots of three different sets of data that have been transformed into Z scores.

Set 1

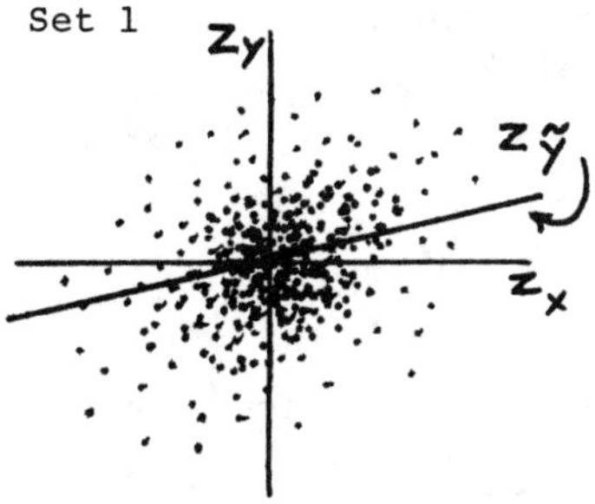

Notice that the relationship between Z_x and Z_y is weak, thus the angle $\theta°$ is only 3° and tan $\theta°$ = r = .05

Set 2

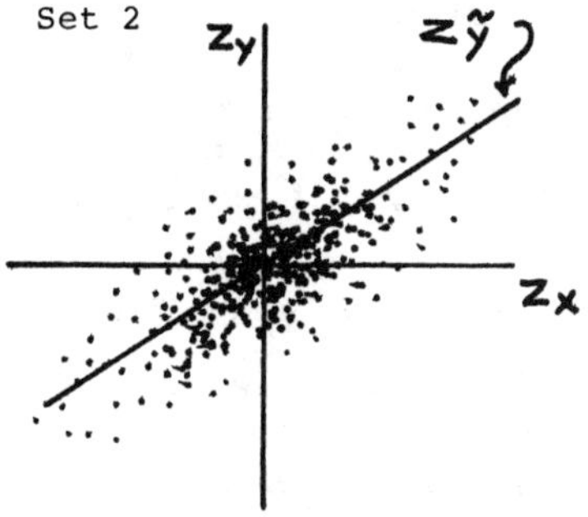

Notice that there is a moderate positive relationship between Z_x and Z_y. Thus, the angle $\theta°$ is greater here than for the data that make up set 1. In the present instance $\theta°$ = 35° and the tan 35° = r = .70.

Set 3

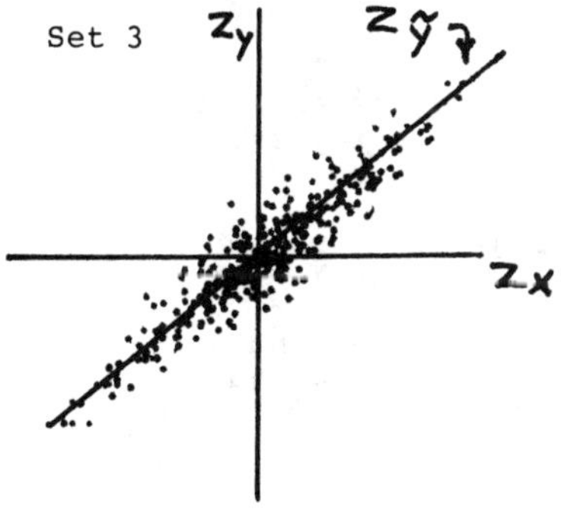

Notice that there is a relatively strong positive relationship between Z_x and Z_y. For these data the angle $\theta°$ = 43° and the tan 43° = r = .93.

In considering the scatterplots on the previous page it is important to recognize that in most instances the values of b and a in the expression $\tilde{Y} = bX + a$ will not be the same as the values of b and a in the expression $z_{\tilde{y}} = bz_x + a$ Indeed it is only when our data are expressed in Z score form that the value of a is always zero and the value of b is always equal to

$$r = \frac{\Sigma z_x z_y}{N}$$

We suggested earlier that $r^2 = 1 - \frac{\sigma^2_{err}}{\sigma^2_y}$ where

σ^2_{err} is the variance around the regression line.

By now it should be apparent that we could, if we wanted to, also express r^2 as equal to

$$1 - \frac{\sigma^2_{err\ zy}}{\sigma^2_{zy}}$$

or since $\sigma^2_{zy} = 1$, we can assert that $r^2 = 1 - \sigma^2_{err\ zy}$

It may not be obvious but we also have the option of expressing r^2 in either of the following ways:

(1) $r^2 = \frac{\sigma^2_{\tilde{y}}}{\sigma^2_y}$ or (2) $r^2 = \sigma^2_{z\tilde{y}}$

The first formula asserts that r^2 is equal to $\sigma^2_{\tilde{y}}$ (the variance of the points on the regression line) divided by σ^2_y (the variance of the Y scores).

If we are to calculate $\sigma^2_{\tilde{y}}$ for a given set of data, we must use the regression equation $\tilde{Y} = bX + a$ to determine the value of $\tilde{Y}$ for each value of X. Thus, if there are N items (each with an X and a Y value) there will be N values of $\tilde{Y} = bX + a$ and $\sigma^2_{\tilde{y}}$ (the variance of this set of values) will have the following form

$$\sigma^2_{\tilde{y}} = \frac{\Sigma(\tilde{y} - \frac{\Sigma\tilde{y}}{N})^2}{N}$$

Let's look first at the expression $\frac{\Sigma\tilde{y}}{N}$. This is the mean of the predicted values of Y for each value of X and for a given set of items $\frac{\Sigma\tilde{y}}{N}$ will always equal $\frac{\Sigma Y}{N} = \mu_y$

This numerical equivalence occurs because each $\tilde{Y}$ is equal to bX + a, hence we know that

$$\Sigma\tilde{y} = \Sigma(bX + a)$$

Thus

$$\Sigma\tilde{y} = b\Sigma X + Na$$

We also know that the values of b and a in the above equation were derived by solving the simultaneous equations:

$$a\Sigma X + b\Sigma X^2 = \Sigma XY$$

and

$$aN + b\Sigma X = \Sigma Y$$

Notice that the second of these simultaneous equations asserts that a and b must be such that

$$\Sigma y = b\Sigma X + aN$$

but if

$$\Sigma\tilde{y} = b\Sigma X + aN$$

we can be sure that $\Sigma\tilde{y} = \Sigma y$

and hence that $\dfrac{\sum \tilde{Y}}{N} = \dfrac{\sum Y}{N}$

Accordingly, we can assert that

$$\sigma^2_{\tilde{Y}} = \frac{\sum(\tilde{Y} - \mu_Y)^2}{N}$$

The importance of this equivalence becomes apparent when we learn that for a given set of items (each with an X and a Y measure) the variance of the Y values (σ^2_Y) can be expressed as the sum of two variances such that

$$\sigma_Y^{\,2} = \sigma_{err}^{\,2} + \sigma^2_{\tilde{Y}}$$

In other words, for a given set of items

$$\frac{\sum(Y-\mu_Y)^2}{N} = \frac{\sum(Y-\tilde{Y})^2}{N} + \frac{\sum(\tilde{Y}-\mu_Y)^2}{N}$$

We can see why if we consider that $\sigma_Y^{\,2}$ can be written as $\dfrac{\sum[(Y-\tilde{Y})+(\tilde{Y}-\mu_Y)]^2}{N}$

Here we have added and subtracted Y to each value of $(Y - \mu_Y)$

Of course the value of $\tilde{Y}$ is different for each value of X (and hence Y) but for a given value of Y we add and subtract the same value of $\tilde{Y}$ and hence we do not change the value of the expression.

If we now square the terms in the brackets we get

$$\sum[(y-\tilde{y}) + (\tilde{y}-\mu_y)]^2$$

$$= \sum(y-\tilde{y})^2 + 2\sum(y-\tilde{y})(\tilde{y}-\mu_y)$$

$$+ \sum(\tilde{y}-\mu_y)^2$$

For now let's focus on the middle term:

$$2\sum(y-\tilde{y})(\tilde{y}-\mu_y)$$

If we can prove that for any set of items this term is equal to zero it will be an easy matter to show that for any set of items

$$\sigma_y^2 = \sigma_{err}^2 + \sigma_{\tilde{y}}^2$$

Here is a proof that the middle term is equal to zero.

$$\sum(y-\tilde{y})(\tilde{y}-\mu_y)$$

$$= \sum[y\tilde{y} - y\mu_y - \tilde{y}\tilde{y} + \mu_y\tilde{y}]$$

$$= \sum y\tilde{y} - \mu_y\sum y - \sum\tilde{y}\tilde{y} + \mu_y\sum\tilde{y}$$

but we already showed that for a given set of items $\sum y = \sum\tilde{y}$ hence the expression becomes

$$\sum y\tilde{y} - \mu_y\sum y - \sum\tilde{y}\tilde{y} + \mu_y\sum y$$

$$= \sum y\tilde{y} - \sum\tilde{y}\tilde{y}$$

Our problem now is to show that $\sum y\tilde{y} = \sum\tilde{y}\tilde{y}$

and hence that $\sum y\tilde{y} - \sum \tilde{y}\tilde{y} = 0$

Let's consider one term at a time. First we'll deal with $\sum y\tilde{y}$.

Since $$\tilde{y} = bx + a$$

$$\sum y\tilde{y} = \sum (y[bx + a])$$
$$= \sum ybx + \sum ya$$
$$= b\sum xy + a\sum y$$

but from the simultaneous equations we know that

$$\left[a\sum x + b\sum x^2\right] = \sum xy$$

thus $$\sum y\tilde{y} = b\left[a\sum x + b\sum x^2\right] + a\sum y$$
$$= \left[ab\sum x + b^2\sum x^2\right] + a\sum y$$

Since the simultaneous equations also tell us that

$$aN + b\sum x = \sum y$$

we can express $a\sum y$ as $a^2N + ab\sum x$

In other words we can assert that

$$\sum y\tilde{y} = ab\sum x + b^2\sum x^2 + a^2N + ab\sum x$$
$$= b^2 + 2ab\sum x + a^2N$$

Now let's consider the term $\sum \tilde{y}\tilde{y}$

Since $$\tilde{y} = bx + a$$

$$\tilde{y}\tilde{y} = [bx + a][bx + a]$$

$$= b^2x^2 + 2abx + a^2$$

and $\sum \tilde{y}\tilde{y} = b^2\sum x^2 + 2ab\sum x + a^2N$

As seen here $\sum y\tilde{y}$ and $\sum \tilde{y}\tilde{y}$ are both equal to $b^2\sum x^2 + 2ab\sum x + a^2N$

thus it must be true that $\sum y\tilde{y} = \sum \tilde{y}\tilde{y}$

Since, however,

$$\sum(y-\tilde{y})(\tilde{y}-\mu_y) = \sum y\tilde{y} - \sum \tilde{y}\tilde{y}$$

the quantity $\sum(y-\tilde{y})(\tilde{y}-\mu_y)$ must equal zero. All of this means that for any set of bivariat data

$$\sum(y-\mu_y)^2 = \sum(y-\tilde{y})^2 + \sum(\tilde{y}-\mu_y)^2$$

Let's divide both sides of this equation by N. When we do so we obtain

$$\frac{\sum(y-\mu_y)^2}{N} = \frac{\sum(y-\tilde{y})^2}{N} + \frac{\sum(\tilde{y}-\mu_y)^2}{N}$$

or as asserted earlier

$$\sigma_y^2 = \sigma_{err}^2 + \sigma_{\tilde{y}}^2$$

If we divide both sides of this equation by σ_y^2 we discover that

$$\frac{\sigma_y^2}{\sigma_y^2} = \frac{\sigma_{err}^2}{\sigma_y^2} + \frac{\sigma_{\tilde{y}}^2}{\sigma_y^2}$$

When we rearrange this expression we find that

$$\frac{\sigma^2_{err}}{\sigma^2_Y} + \frac{\sigma^2_{\tilde{Y}}}{\sigma^2_Y} = 1 \quad \text{thus} \quad \frac{\sigma^2_{\tilde{Y}}}{\sigma^2_Y} = 1 - \frac{\sigma^2_{err}}{\sigma^2_Y}$$

Since, however $r^2 = 1 - \dfrac{\sigma^2_{err}}{\sigma^2_Y}$

We can conclude that $r^2 = \dfrac{\sigma^2_{\tilde{Y}}}{\sigma^2_Y}$

Moreover, if we transform our X s and Y s into z_X s and z_Ys, we find that since $\sigma^2_{z_Y} = 1$

$$r^2 = \frac{\sigma^2_{z\tilde{Y}}}{\sigma^2_{z_Y}} = \sigma^2_{z\tilde{Y}}$$

In addition to providing an alternative way of quantifying the relationship between X and Y, the expression

$$r^2 = \frac{\sigma^2_{\tilde{Y}}}{\sigma^2_Y}$$

provides us with an important insight into the meaning of r .

In essence $r = \dfrac{\sigma^2_{\tilde{Y}}}{\sigma^2_Y}$ says that the square of r

is the proportion of the variance of our Y measures that we can account for through knowledge of X. Stated somewhat differently, the quantity r^2 tells us how much of the variance of Y is accounted for by the variance of the estimate values of Y for each value of X.

When $r = 0$, $r^2 = 0$, hence we cannot account for any part of the variance of Y through our knowledge of X.

When, on the other hand, $r = 1$, $r^2 = 1$, which means that we can account for all of the variation in Y through our knowledge of X.

As was seen above, we arrived at these conclusions by making use of the proposition that

$$\sigma_y^2 = \sigma_{err}^2 + \sigma_{\tilde{y}}^2$$

The material on the next few pages illustrates certain of these key ideas graphically.

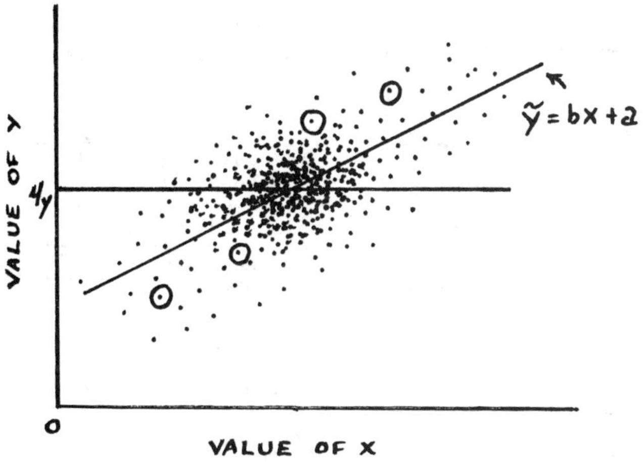

We have circled four of the items in this scatterplot in order to illustrate the deviations that play a role in the expression $\sigma_y^2 = \sigma_{err}^2 + \sigma_{\tilde{y}}^2$.

Of course we have circled only four of the N items that must be considered when we calculate the parameters (a & b) of the regression line, but with some thought it should be apparent that if we can understand what happens with respect to these four items we can understand what happens with respect to all N items.

Here are three scatterplots showing only the four circled items from the scatterplot on the previous page. These scatterplots again show the regression line and the line indicating the mean of the Y values. Also shown in these scatterplots are the deviations that are employed in the calculations of σ_Y^2, σ_{err}^2 + $\sigma_{\tilde{Y}}^2$

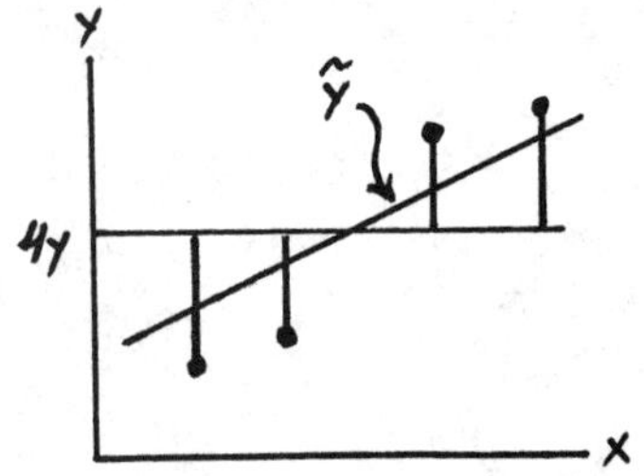

The quantity (σ_Y^2) is is the average of the squares of these deviations, i.e.

$$\sigma_Y^2 = \frac{\sum (Y-\mu_Y)^2}{N}$$

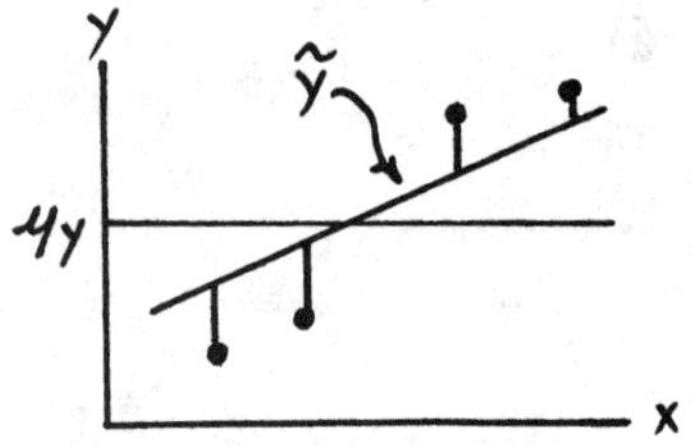

The quantity (σ_{err}^2) is the average of the squares of these deviations, i.e.

$$\sigma_{err}^2 = \frac{\sum (Y-\tilde{Y})^2}{N}$$

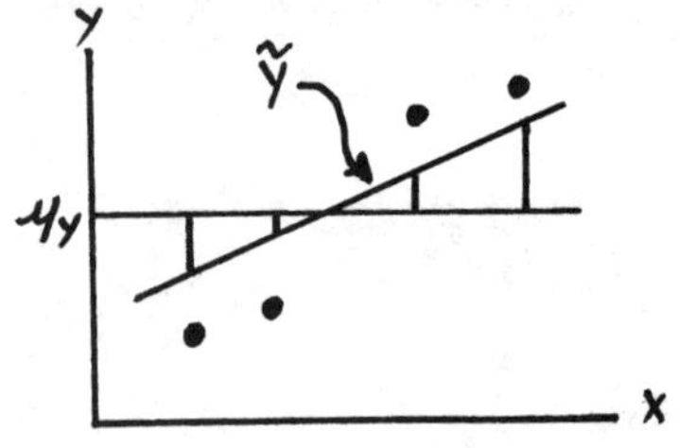

The quantity ($\sigma_{\tilde{Y}}^2$) is the average of the squares of these deviations, i.e.

$$\sigma_{\tilde{Y}}^2 = \frac{\sum (\tilde{Y}-\mu_Y)^2}{N}$$

As seen in these scatterplots, each of the deviations that go into σ_Y^2 can be conceived to consist of the sum of two other kinds of deviations. (1) the deviation of Y from $\tilde{Y}$ and (2) the deviation of $\tilde{Y}$ from μ_Y. Of course this insight does not itself prove that for any set of items $\sigma_Y^2 = \sigma_{err}^2 + \sigma_{\tilde{Y}}^2$ but we provided an algebraic proof of this proposition on pages 190-193. Once we proved this proposition we were able to show that if $r^2 = 1 - \frac{\sigma_{err}^2}{\sigma_Y^2}$ it must also be true that $r^2 = \frac{\sigma_{\tilde{Y}}^2}{\sigma_Y^2}$ and if our data are transformed into Z scores, it must be true that

$$r^2 = \sigma_{Z_{\tilde{Y}}}^2$$

Earlier (on page 177) it was stated that we would show that

$$r^2 = 1 - \frac{\sigma_{err}^2}{\sigma_Y^2}$$

While we have shown this numerical equivalence for a particular set of items (i.e., for the data on page 174) we have not yet demonstrated the truth of the general proposition that for any set of items

$$r^2 = 1 - \frac{\sigma_{err}^2}{\sigma_Y^2}$$

We can now show this by pointing out that if the X and Y values of our items are transformed into Z scores the above expression becomes

$$r^2 = 1 - \sigma_{err_{ZY}}^2 = \sigma_{Z_{\tilde{Y}}}^2$$

i.e.,

$$r^2 = \sigma_{Z_{\tilde{Y}}}^2$$

Since, however, $Z_{\tilde{Y}} = bZ_X + a$ and since we showed that when our data are in Z score form $b = \frac{\Sigma Z_X Z_Y}{N}$

and a = 0.

We know that $z_{\tilde{y}} = \left[\frac{\Sigma z_x z_y}{N}\right] z_x$

We also know that the variance of the z_x's (i.e., $\sigma^2_{z_x}$) is one.

Accordingly, we can be sure that if we multiply every z_x by a constant (in this case $\frac{\Sigma z_x z_y}{N}$ we multiply the variance by the square of the constant

Thus if $\sigma^2_{z_x} = 1$

$$\sigma^2_{\left[\frac{\Sigma z_x z_y}{N}\right] z_x} = \left[\frac{\Sigma z_x z_y}{N}\right](1)$$

In other words, if we assert that $r^2 = \sigma^2_{z_{\tilde{y}}}$

we discover that

$$r^2 = \left[\frac{\Sigma z_x z_y}{N}\right]^2$$

Thus we see that when we approach the concept of correlation through the concepts of regression we arrive at the same expression as when we <u>defined</u> r as

$$r = \frac{\Sigma z_x z_y}{N}$$

Thus far we have talked as if for a correlation problem there were only one regression line. Actually, there is a second regression line that could also be drawn.

The next page shows a scatterplot of the X and Y values for a large number of items. Once again we have circled four of the N items, but this time we have drawn the regression line that would enable us to engage in the procedure of predicting a given item's X value, through knowledge of its Y value. We have also shown (with a dotted line) the regression line used to predict Y from X.

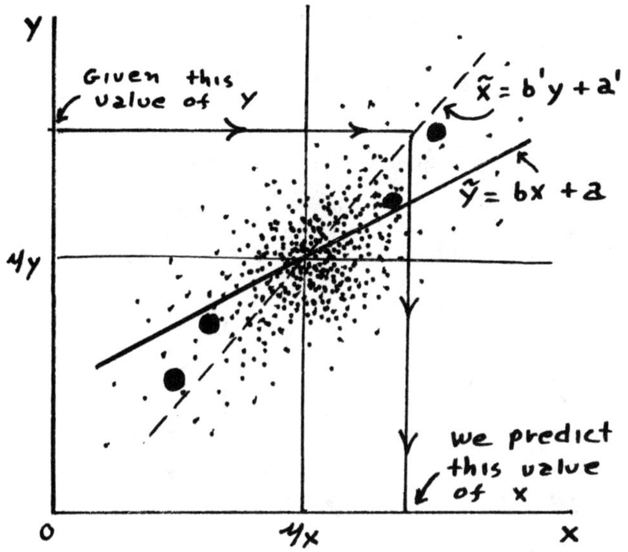

In this figure, the line specified by the equation $\tilde{X} = b'Y + a'$ is called the regression of X on Y. With this terminology the line specified by the equation $\tilde{Y} = bX + a$ (that we studied earlier) is called the regression of Y on X.

We saw earlier that the values of a and b in the regression equation $\tilde{Y} = bX + a$ can be obtained by solving the simultaneous equations

$$a\Sigma X + b\Sigma X^2 = \Sigma XY$$

$$Na + b\Sigma X = \Sigma Y$$

The simultaneous equations that specify the values of a' and b' in the regression equation $\tilde{X} = b'Y + a'$ have, as might be expected, a similar form.

$$a' \sum y + b' \sum y^2 = \sum xy$$
$$N a' + b' \sum y = \sum x$$

The next pair of figures show, for the four emphasized points in the previous scatterplot 1) the deviations that are minimized by the regression line Y = bX + a (on the left), 2) the deviations that are minimized by the regression line X = b Y + a (on the right)

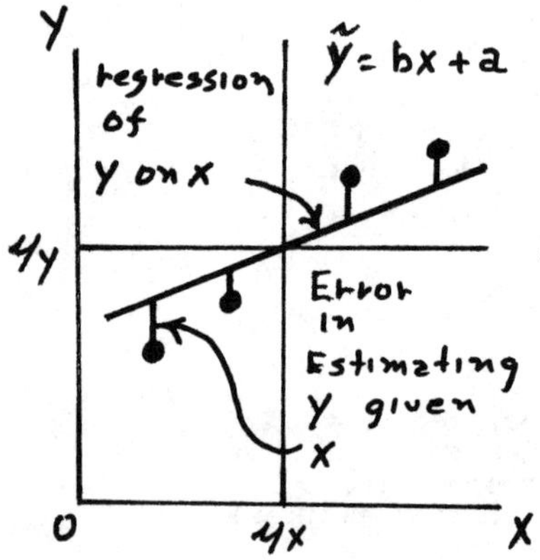

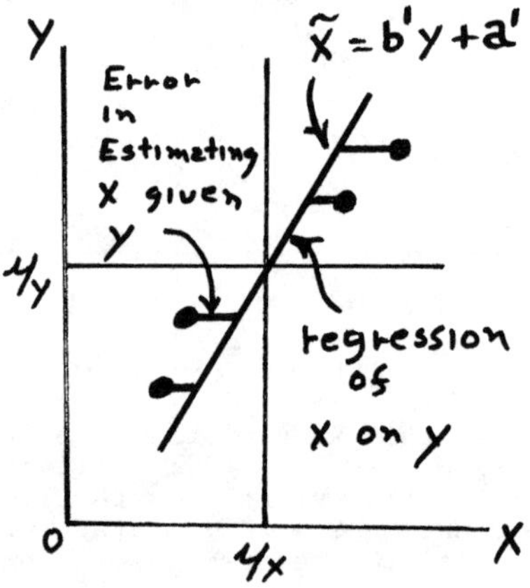

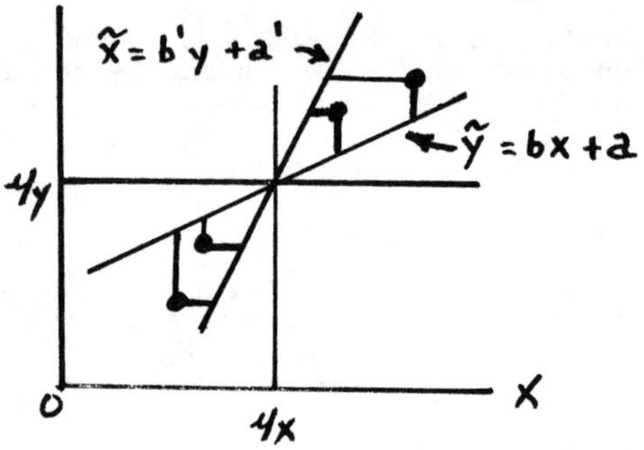

Here are several sets of data plotted in Z score form. You can gain insight into the material we are studying by observing how the regression lines change as the relationship becomes stronger.

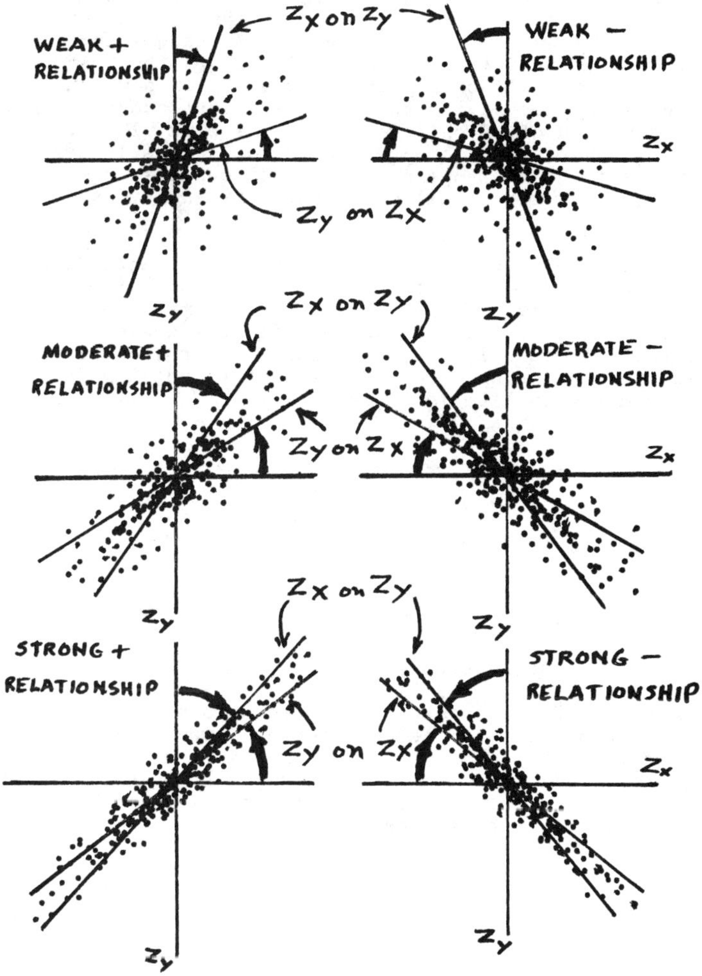

As implied in this illustration, when there is no relationship between the Zx and Zy measures (i.e., r = 0) the two regression lines fall along the two axes. This is as it should be. After all, if there is no relationship between Zx and Zy, for a gven item our best estimate of Zy given Zx will always be the mean of the Zy values. But the mean of the Zy values is zero. This means that for every value of Zx the best estimate of Zy will be a point on the Zx axis which is another way of saying that the line that expresses the regression of Zy on Zx will fall exactly along the Zx axis. Since a similar argument holds for estimates of Zx based on knowledge of Zy, it follows that the two regression lines will overlap the Zx and Zy axis when r = 0.

As the relationship grows stronger the regression lines increasingly depart from the Zx and Zy axis until, for a perfect relationship, they come together (i.e., overlap each other). If one studies these figures and carefully reasons through their logical basis it should always be a simple matter to identify which regression line is which.

Regression toward the mean

One of the most interesting features of correlational data is the phenomenon of regression towards the mean. Consider, for example, the hypothetical data shown on the next page. While these are fictional data they have a form that is very close to what usually occurs when heights in families are measured.

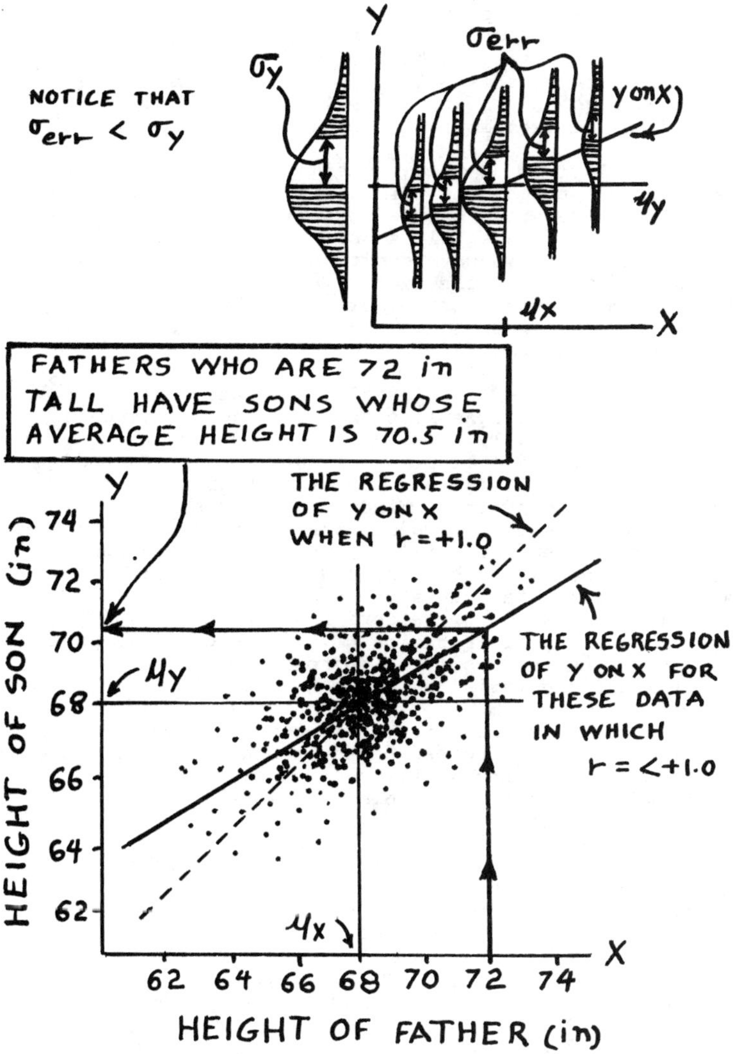
NOTICE THAT
σerr < σy
σy
Y
σerr
y on X
μy
μx
X
FATHERS WHO ARE 72 in TALL HAVE SONS WHOSE AVERAGE HEIGHT IS 70.5 in
THE REGRESSION OF Y ON X WHEN r = +1.0
THE REGRESSION OF Y ON X FOR THESE DATA IN WHICH r = <+1.0
HEIGHT OF SON (in)
HEIGHT OF FATHER (in)
62
64
66
68
70
72
74

In the scatterplot on the previous page, as in every scatterplot of bivariate data, each point represents two measures on a single item. In this instance each item is a family and the two measures are the height of the father and the height of his oldest son at maturity.

The sloping elliptical configuration of these data makes it clear that there is at least a moderate positive correlation between the two measures. Tall fathers tend to have tall sons. Short fathers, on the other hand, tend to have short sons. Moreover, as seen in the scatterplot, the average height of all fathers is the same as the average height of all sons. Note the identical means of the two marginal distributions in the inset on page 203.

As we examine this inset we also see that the two marginal distributions also have the same variance and both are essentially normal. This - and the other features of the scatterplot - is no mere artifact of this particular set of data. Data such as these are obtained with many other kinds of measures. Had we, for example, assessed IQs of fathers and sons in each of a large number of families, the scatterplot would look very much like this one and it would exhibit the same properties that this one exhibits. The distribution of IQs of fathers would be essentially the same as the distribution of the IQs of their sons. Both distributions would be approximately normal and their means and variances would be virtually identical. It is important to emphasize this point in order to make it clear that the phenomenon we are about to examine (regression toward the mean) occurs even when, as in the scatterplot on page 203, the means of the two marginal distributions are exactly equal to each other.

To observe regression toward the mean we need only examine the scatterplot on page 203 and ask "What is the average height of those sons whose fathers' height is 72 inches?" When we examine the scatterplot carefully, we arrive at the disconcerting realization that the answer is approximately 70.5 inches. In other words, the average height of sons with fathers whose heights are 4 inches above the mean of all heights is itself only 2.5 inches above the mean of all heights. In statistical terminology, the heights of sons exhibit regression toward the mean. This is a general property of any scatterplot where the linear relationship is less than perfect. Under such circumstances the mean of the Y measures for a given value of X will be closer to the mean of all the Y measures than the X measure is to the mean of its distribution. In a similar fashion the mean of the X measures for a given value of Y will be closer to the mean of the X measures than the Y measure is to the mean of its distribution. If the correlation coefficient were zero, the mean of the Y measures for each value of X would always be the mean of all the Y values and visa versa, i.e., we would find complete regression toward the mean. If, on the other hand, the correlation were perfect (r = either ± 1) the mean of the Y values for each value of X would be as far from the mean of the Y values as the X value is from the mean of its distribution. In other words there would be no regression towards the mean. This leads to the conclusion that the phenomenon of regression toward the mean makes its appearance when we deal with a less than perfect linear relationship such as the one illustrated on page 203.

As seen on page 203, the line that expresses the regression of Y on X tends to pass through the means of the heights of sons whose fathers have the heights indicated on the X axis. This fact is also illustrated in the inset in the upper right corner which portrays the distributions of Y values for a variety of values of X. As seen there, for a given X value, the Y values are normally distributed and the mean of each distribution falls on the line that expresses the regression of Y on X. While it may not be apparent at first, aside from their different means, the visual differences between these distributions are primarily due to the fact that each has a different number of cases. Thus the distributions in the middle of the cluster have many more cases than those at the borders, but as noted above, and for much of the data provided by nature, all of these distributions tend to be normal. Moreover, when actual measurements are taken all of these distributions tend to have the same variance. In general, the quantity σ^2_{err} is a measure of this variance. This statement will begin to make sense if one considers it in relation to the way in which we originally defined the coefficient of correlation.

$$r^2 = 1 - \frac{\sigma^2_{err}}{\sigma^2_y}$$

In this equation σ^2_{err} is the variance around the regression line. Numerically it is the average of all of the squared deviations around the regression line. If, for each value of X, the Y values are approximately equally scattered about the regression line we have the condition of equal variances described above.

It is perhaps of interest that some students have been known to greatly impress their friends by casually mentioning that this condition of equal variances is described by the technical term "homoscedasticity".

There is one more important point to make about regression toward the mean. We have just seen that when heights in families are measured, tall parents tend to have tall children but these children are not, on the average, as tall as their parents. It is also, of course true that (as seen on page 203) parents with heights below average tend to have children whose heights are also below average but, in general, these children are somewhat taller than their parents.

When they first encounter this fact, some students are tempted to conclude that the human race must be inexorably moving toward equality of heights. After all, if tall parents tend to have children who are shorter than themselves and if short parents tend to have somewhat taller children, given enough successive generations, one ought to eventually wind up with a population in which everyone's height is exactly 68 inches.

Fortunately, however, this conclusion is in error. It neglects the important fact that in a given generation, while many of the very tall sons come from tall parents, many also come from parents with heights that are at or even below the average. (If you carefully examine the scatterplot on page 203 you can verify this for yourself.) Tall children are rarer in families where the parents have near average height than in families where the parents are tall, but there are so many more parents whose height is near average than parents who are very tall, that in successive generations any losses of tall children (as reflected in regression toward the mean) can be counteracted by the supply of tall children from that large group of parents whose height happens to be near average. In other words, while a large proportion of the tall children in a given generation may be the offspring of parents who are tall, a sizeable proportion of those children also come from parents whose heights are near average.

Some Final Considerations with Respect to Correlation and Regression

On page 164 it was asserted that the methods outlined here were developed to deal with problems of correlation and regression for data in which the relationship is linear. (Recall that in a linear relationship the data tend to cluster along a straight line.) The thoughtful student will recognize, however, that one could apply the present methods to any set of data so long as one could legitimately identify an X and a Y measure on each of the set of N items. That student will also recognize that if these methods were used on data that yielded a nonlinear cluster (look again at p. 164), the procedures would only provide an index of whatever linear relationship might exist in the data.

Finally, it is also important to note that when a given set of bivariate data exhibit a relationship (linear or otherwise) this fact does not itself imply that either measure is the cause of the other. From all that we have said thus far it should be clear that the correlation coefficient is merely a number that provides a quantitative index of the linear relationship in the data but that number tells us nothing about why that relationship was observed. Certainly there are several possible reasons. For example, the obtained r might reflect a chance occurrence.

The next page shows a large population of items in which the X and Y measures exhibit a zero relationship.

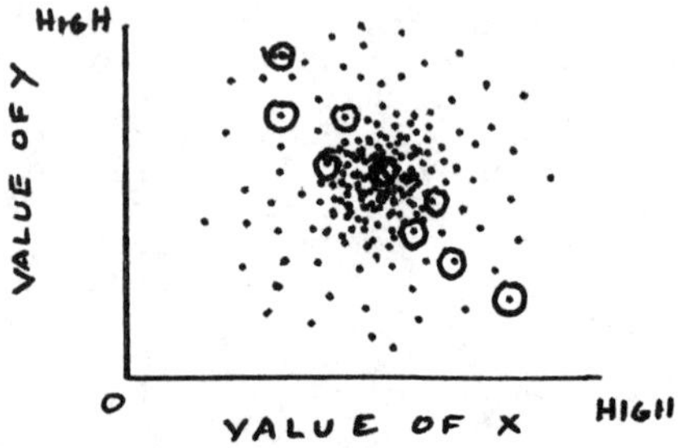

Suppose that we take a random sample of items from this population and happen to pick the particular ones that are circled. Those items would yield a calculated r equal to perhaps -.80 even though the r for the parent population is actually zero. Of course, there is a test for whether or not a given r is among those rare events that would occur only, say, one out of twenty times when we repeatedly take random samples from a population where the r is equal to zero. While presentation of this test (The Fisher r to Z transformation) is beyond the scope of the present work, there are a number of clear accounts available in most intermediate and in some introductory textbooks.

But even if an obtained r could be shown to be statistically significant, this fact should not be taken to mean that one variable bears a causal relationship to the other. It is possible, for example, that the variation in both measures is produced by a third (unassessed) or uncontrolled variable. For example, it seems apparent that the length of a person's arm is unrelated to the size of their vocabulary. But if we took a sample of children without regard to age, we would undoubtedly obtain a high positive correlation on these two measures. This is because both vocabulary and arm length in a developing child is related to age. Very young children have short arms and they also have small vocabularies. Older children, on the other hand, have much longer arms and reasonably large vocabularies. Clearly, when we deal with statistics, there is no ready substitute for a thoughtful and knowledgeable approach.

Chapter 12

TESTS OF HYPOTHESES ABOUT DIFFERENCES BETWEEN POPULATION MEANS WHEN THE MEASURES IN THE TWO SAMPLES ARE RELATED

Earlier (on pages 129 through 139) we saw how we could use the statistic

$$z_{(\bar{X}_1-\bar{X}_2)} = \frac{(\bar{X}_1 - \bar{X}_2) - (\mu_{h_1} - \mu_{h_2})}{\sqrt{\dfrac{\sigma_{X_1}^2}{n_1} + \dfrac{\sigma_{X_2}^2}{n_2}}}$$

to test the statistical hypothesis that the samples came from populations where $\mu_{X_1} - \mu_{X_2} = \mu_{h_1} - \mu_{h_2}$. One of the assumptions of this test is that the measures in each of the two samples are randomly selected and while we did not mention it earlier, this implies that the measures in the two samples are independent of each other. There are occasions, however, where this latter condition does not obtain and, instead of dealing with two sets of independent measures, we find that measures in one sample bear a statistical relationship to the measures in the other.

For instance, if we wanted to examine the relative efficiency of two different teaching methods we might elect to form matched classes, where each subject in one class had a counterpart in the other whose IQ was similar. Under these circumstances the subjects in one of the classes might have been randomly selected, but the necessity to match IQs would preclude random selection of subjects for the second class. Another similar procedure (which also controls possible differences in mean IQ across classes) would be to expose the same set of subjects to both teaching methods. With either of these designs (as when we test two randomly selected groups) we would obtain two sets of scores but, unlike a fully random design, the matching technique would lead us to expect that the scores in one condition would be related to the scores in the other. Under such circumstances it would be inappropriate to treat the two sets of measures as independent random samples and hence it would be inappropriate to use the statistic

$Z_{(\bar{X}_1 - \bar{X}_2)}$ to test the hypothesis that $\mu_{X_1} - \mu_{X_2} = 0$

How then can we decide whether or not the two teaching methods have produced reliably different effects? For reasons that will become clear later we can get at this issue if we rephrase our statistical question so that, instead of asking whether the observed difference between means is statistically reliable, we ask whether the mean of the observed differences (measured across subjects) is significantly different from zero.

When stated in this fashion our question reduces to a test of an exact hypothesis about the mean of a single population.

On page 69 we saw how we could use the statistic

$$Z_{\bar{X}} = \frac{\bar{X} - \mu_h}{\sqrt{\frac{\sigma_x^2}{n}}}$$

to carry out such a test.

The procedure that we use when we want to test for differences between related measures is a variation of this technique. Here, for example, is a set of n items with two measures on each.

Item No.	X measure	Y Measure	D = X - Y
1	X_1	Y_1	$D_1 = X_1 - Y_1$
2	X_2	Y_2	$D_2 = X_2 - Y_2$
⋮	⋮	⋮	⋮
n	X_n	Y_n	$D_n = X_n - Y_n$

To make things concrete, imagine that the items are people and that the X measure is a given individual's grade under teaching method A, whereas the Y measure is the same individual's grade under method B.

In terms of the above table, the several Ds would represent the set of observed differences in grades across subjects. Our problem is to decide whether the mean of these Ds is among those extreme cases that occur only, say, one out of 20 times by chance when we repeatedly draw samples of size n from a population where $\mu_D = 0$. In other words, we are asking whether our obtained $\bar{D} = \frac{\Sigma D}{n}$ is sufficiently different from zero to enable us to reject the statistical hypothesis that the mean of the population from which it was obtained is itself zero.

If this hypothesis is true and if the population of Ds is normally distributed (or if n is large), the statistic

$$Z_{\bar{D}} = \frac{\bar{D} - \mu_h}{\sqrt{\frac{\sigma_D^2}{n}}} \quad \left\{ \begin{array}{l} \text{Here} \\ \mu_h = 0 \end{array} \right.$$

will have a normal sampling distribution with mean = 0 and variance = 1.

Of course the use of this statistic (i.e., $Z_{\bar{D}}$) presupposes that we know the value of σ_D^2 , but as noted earlier in this book, knowledge of the value of the variance is always a prerequisite for the use of the statistic Z.

Perhaps the best way to gain insight into the special features of the $Z_{\bar{D}}$ test is to rewrite its formula in terms of Xs and Ys instead of Ds. As we do so, it will be helpful to refer back to the set of items on the bottom of page 210 where we see that for a given item D is always equal to X - Y.

The first point to recognize is that when rewritten in terms of Xs and Ys the $\bar{D}$ is numerically equal to $(\bar{X} - \bar{Y})$ This follows because $\bar{D} = \frac{\sum D}{n}$

thus $\bar{D} = \frac{\sum(X - Y)}{n} \quad \frac{\sum X - \sum Y}{n}$

$$= \frac{\sum X}{n} - \frac{\sum Y}{n} = \bar{X} - \bar{Y}$$

The next point to recognize is that the quantity $\sigma_{\bar{D}}^2$ that appears in the denominator of the formula for $Z_{\bar{D}}$ is actually much more complex than might appear at first glance.

We begin to appreciate this fact when we attempt to rewrite the equation that defines σ_D^2 in terms of Xs and Ys.

Here is the formula for σ_D^2

$$\sigma_D^2 = \frac{\sum(D - \mu_D)^2}{N}$$

Note that N appears here because σ_D^2 is the variance of the population of Ds.

$$\sigma_D^2 = \frac{\sum[(X - Y) - (\mu_X - \mu_Y)]^2}{N}$$

We just showed that D=X-Y. By a similar logic it must also be true that $\mu_D = \mu_X - \mu_Y$

$$\sigma_D^2 = \frac{\sum[X - Y - \mu_X + \mu_Y]^2}{N}$$

$$\sigma_D^2 = \frac{\Sigma[(X - \mu_X) - (Y - \mu_Y)]^2}{N}$$

$$\sigma_D^2 = \frac{\Sigma[(X - \mu_X)^2 - 2(X - \mu_X)(Y - \mu_Y) + (Y - \mu_Y)^2]}{N}$$

$$= \frac{\Sigma(X - \mu_X)^2}{N} - 2\frac{\Sigma(X - \mu_X)(Y - \mu_Y)}{N} + \frac{\Sigma(Y - \mu_Y)^2}{N}$$

$$= \sigma_X^2 - 2\frac{\Sigma(X - \mu_X)(Y - \mu_Y)}{N} + \sigma_Y^2$$

$$= \sigma_X^2 + \sigma_Y^2 - 2\frac{\Sigma(X - \mu_X)(Y - \mu_Y)}{N}$$

As we shall see in a moment, the quantity

$$\frac{\Sigma(X - \mu_X)(Y - \mu_Y)}{N}$$

can be shown to be equal to $r\sigma_X\sigma_Y$. Hence

$$\sigma_D^2 = \sigma_X^2 + \sigma_Y^2 - 2r\sigma_X\sigma_Y$$

To see why this is so we need to recall that for a given population of items with an X and a Y measure on each

$$r = \frac{\Sigma z_X z_Y}{N}$$

where

$$z_x = \frac{X - \mu_x}{\sigma_x} \quad \text{and} \quad z_y = \frac{Y - \mu_y}{\sigma_y}$$

and N refers to the number of items in the population

This means that

$$r = \frac{\sum\left(\frac{X - \mu_x}{\sigma_x}\right)\left(\frac{Y - \mu_y}{\sigma_y}\right)}{N}$$

$$r = \frac{\sum (X - \mu_x)(Y - \mu_y)}{N\ \sigma_x\ \sigma_y}$$

$$r\sigma_x \sigma_y = \frac{\sum (X - \mu_x)(Y - \mu_y)}{N}$$

In short, we have just shown that

$$\sigma_D^2 = \sigma_x^2 + \sigma_y^2 - 2r\sigma_x \sigma_y$$

All of this means that when rewritten in terms of Xs and Ys the formula for $z_{\bar{D}}$ is transformed as follows:

$$z_{\bar{D}} = \frac{\bar{D} - \mu_{h_0}}{\sqrt{\frac{\sigma_D^2}{n}}} = \frac{(\bar{X} - \bar{Y}) - (\mu_{h_x} - \mu_{h_y})}{\sqrt{\frac{\sigma_x^2 + \sigma_y^2 - 2r\sigma_x\sigma_y}{n}}}$$

$$z_{\bar{D}} = \frac{(\bar{X} - \bar{Y}) - (\mu_{h_x} - \mu_{h_y})}{\sqrt{\frac{\sigma_x^2}{n} + \frac{\sigma_y^2}{n} - 2r\sigma_x\sigma_y\left(\frac{1}{n}\right)}}$$

Here are the two formulas for Z, side by side:

$$Z_{\bar{D}} = \frac{\bar{D} - \mu_D}{\sqrt{\frac{\sigma_{\bar{D}}^2}{n}}} \qquad Z_{(\bar{X}-\bar{Y})} = \frac{(\bar{X} - \bar{Y}) - (\mu_X - \mu_Y)}{\sqrt{\frac{\sigma_X^2}{n} + \frac{\sigma_Y^2}{n} - 2r\frac{\sigma_X \sigma_Y}{n}}}$$

Since, as we have just seen, one is the algebraic transformation of the other, for a given set of data they will be numerically equivalent.

We can see some of the special features of these formula if we compare them to the formula for $Z_{(\bar{X}_1 - \bar{X}_2)}$ that we use when we are testing a statistical hypothesis about population means based on measures that are independent of each other.

Here is that formula

$$Z_{(\bar{X}_1 - \bar{X}_2)} = \frac{(\bar{X}_1 - \bar{X}_2) - (\mu_1 - \mu_2)}{\sqrt{\frac{\sigma_{X_1}^2}{n} + \frac{\sigma_{X_2}^2}{n}}}$$

As can be seen, the formulas for $Z_{(\bar{X}-\bar{Y})}$ and $Z_{(\bar{X}_1 - \bar{X}_2)}$ are very similar; the difference being that when our measures are independent the last term under the radical in the denominator is missing. To see why, we need recall that when we assert that the items in the two samples are randomly selected we are also asserting that the correlation between them is zero. Notice that if r = 0 the two formula become numerically identical.

This is a very interesting set of circumstances. It means that if the r in a given test of related measures is positive we will obtain a larger value of

Z on a given set of data than we would have obtained if we treated the items as if they were independent measures. In other words, if we are dealing with related measures (that is, we have data that can be legitimately described as a set of items with two measures on each), we are more likely to obtain a statistically significant value of Z if the correlation is positive than if it is 0 or if it is negative. In this respect it is important to again emphasize that

$$Z_{\bar{D}} = \frac{\bar{D} - \mu_h}{\sqrt{\dfrac{\sigma_D^2}{n}}} = \frac{(\bar{X} - \bar{Y}) - (\mu_{hx} - \mu_{hy})}{\sqrt{\dfrac{\sigma_X^2}{n} + \dfrac{\sigma_Y^2}{n} - 2r\dfrac{\sigma_X \sigma_Y}{n}}}$$

This algebraic (and hence numerical) equivalence means that when we employ the strategy of testing the mean of differences on a set of related measures we automatically take account of the correlation between the two sets of measures on our items.

Test of Hypotheses about the Means of Two Related Populations When the Values of the Variances are not Known

We noted earlier that in order to use the statistic $Z_{\bar{D}}$ it is necessary to know the value of σ_D^2 in the population. Ordinarily this information would not be available to us, but since $S_X^2 \rightarrow \sigma_X^2$ we could always estimate the value of σ_D^2 by calculating S_D^2 on the set of items before us.

where $S_D^2 = \dfrac{\sum(D - \bar{D})^2}{n - 1}$ with $\bar{D} = \dfrac{\sum D}{n}$

Under such circumstances the appropriate statistic to employ in our test is as follows:

$$t_{(df=n-1)} = \frac{\bar{D} - \mu_{hD}}{\sqrt{\dfrac{s_D^2}{n}}}$$

With this statistic the degrees of freedom is n - 1 where n refers to the number of pairs of items. Here is the formula for t expressed in two ways.

$$t_{(df=n-1)} = \frac{\bar{D} - \mu_{hD}}{\sqrt{\dfrac{s_D^2}{n}}}$$

$$t_{(df=n-1)} = \frac{(\bar{X} - \bar{Y}) - (\mu_{hx} - \mu_{hy})}{\sqrt{\dfrac{S_X^2}{n} + \dfrac{S_Y^2}{n} - 2r' \dfrac{S_X S_Y}{n}}}$$

The formula on the top is the statistic that we would use in a test of a hypothesis about the mean of a set of differences for related measures and the formula on the bottom is the same statistic expressed in terms of Xs and Ys rather than in terms of Ds. Notice that in both cases the number of degrees of freedom is (n - 1) and notice that the formula on the bottom incorporates the values of S_X^2 and S_Y^2 computed on the items in the sample. In this respect it is also important to recognize that the r' that appears in the formula on the bottom is the correlation coefficient for the specific items in the sample before us. In general, the value of r' will not be identical to the r that we might obtain if we were able to measure all items in the population. To understand why this is so it may be helpful to review again the algebraic transformations that were used to express the formula for $z_{\bar{D}}$ in terms of Xs and Ys. With some thought it should become clear that since these transformations retain the numerical equivalence between the two formulas the value of r' in the formula for t must be obtained from the set of items in our sample.

We can perhaps best illustrate the several features of these procedures with a numerical example. Listed below are some numerical measures that might be obtained in a given study. In the one instance we will treat the numbers as representing the data from a study in which we have obtained two measures on each of a set of 5 items (or subjects). In the second instance we will treat the identical numbers as representing the data from a study in which we have obtained one measure on each of two independent samples of 5 items (or subjects) each.

A single sample of n = 5 pairs of related measures

Two independent samples with n = 5 measures in each

X	Y	D	X_1	X_2
3	7	-4	3	7
9	11	-2	9	11
2	4	-2	2	4
5	6	-1	5	6
7	8	-1	7	8
$\bar{X} = 5.20$	$\bar{Y} = 7.20$	$\bar{D} = -2$	$\bar{X}_1 = 5.20$	$\bar{X}_2 = 7.20$
$s_X^2 = 8.18$	$s_Y^2 = 6.71$	$s_D^2 = 1.50$	$s_{X_1}^2 = 8.18$	$s_{X_2}^2 = 6.71$
$s_X = 2.86$	$s_Y = 2.59$	$s_D = 1.22$	$s_{X_1} = 2.86$	$s_{X_2} = 2.59$

$r' = .904$

Shown below are the calculations that would be appropriate for a test of related measures and it compares them to the appropriate calculations for a test of independent measures. Notice that there are two ways that one can do the calculations for related measures.

Related Measures	Related Measures	Independent Measures
$\frac{(\bar{X}-\bar{Y})-(0)}{\sqrt{\frac{s_X^2}{n}+\frac{s_Y^2}{n}-2r'\frac{s_X s_Y}{n}}}$	$\frac{\bar{D}-(0)}{\sqrt{\frac{s_D^2}{n}}}$	$\frac{(\bar{X}_1-\bar{X}_2)-(0)}{\sqrt{\frac{s_{X_1}^2}{n}+\frac{s_{X_2}^2}{n}}}$
$\frac{5.20-7.20-0}{\sqrt{\frac{8.18}{5}+\frac{6.71}{5}-2(.904)\frac{7.41}{5}}}$	$\frac{-2-0}{\sqrt{\frac{1.5}{5}}}$	$\frac{5.20-7.20-0}{\sqrt{\frac{8.18}{5}+\frac{6.71}{5}}}$
$\frac{-2}{\sqrt{1.64+1.34-2.68}}$	$\frac{-2}{\sqrt{.3}}$	$\frac{-2}{\sqrt{1.64+1.34}}$
$\frac{-2}{\sqrt{.3}}$	$\frac{-2}{\sqrt{.3}}$	$\frac{-2}{\sqrt{2.98}}$
$t_{df=4}=-3.65$	$t=-3.65$, $df=4$	$t=-1.16$, $df=8$

As can be seen, the two methods of calculating t for related measures yield values that are identical. Of course we know that the two approaches are algebraically equivalent and hence they should yield exactly the same value of t. The large value of t (i.e., -3.65) obtained in the tests for related measures and the small value of t (i.e., -1.16) obtained in the test for independent measures is an indication of the difference between tests of independent versus related measures. This difference reflects the fact that the test for related measures has a smaller denominator than the test for independent measures.

As seen in this example, for a given observed difference between two means (in this case the difference is -2), we get a larger value of t when our measures are related and the relationship is positive.

Of course, we have fewer dfs in the test for related measures than in the test for independent measures, but as can be seen in Table 1, the t is in the rejection region for $\alpha = .05$ for a two-tailed test of related measures, whereas it falls short of the $\alpha = .05$ rejection region in a two-tailed test for independent measures.

For related measures with $\alpha = .05$ and $df = n - 1 = 5 - 1 = 4$ we need $|t| \geq 2.776$, since we obtained $|t| = 3.65$ we reject the hypothesis tested.

For independent measures with $\alpha = .05$ and $df = n + n - 2 = 5 + 5 - 2 = 8$ we need $|t| \geq 2.306$, since we obtained $|t| = 1.16$ we have no basis to reject the hypothesis tested.

Example VII = Testing a hypothesis about the means of related measures (Variance not known)

Illustration

We wish to determine if reaction times vary systematically as a function of the time of the day. To examine this question, the reaction times of each of a number of randomly selected subjects are tested in the morning and in the afternoon. In order to decide if the mean of the observed differences in reaction times is reliably different from zero we test the statistical hypothesis that this difference represents a random occurrence that happens when samples are drawn from a population of differences where $\mu_D = 0$

Method

(1) Hypothesis: Population of differences has a mean of zero (i.e., $\mu_D = 0$).

(2) Set α

(3) Use $$t_{df = n-1} = \frac{\bar{D} - \mu_h}{\sqrt{\frac{s_D^2}{n}}}$$

n = No. of subjects tested
D = Morning reaction time minus afternoon reaction time for a given subject

$\bar{D} = \frac{\Sigma D}{n}$ = Mean of differences in the sample tested

μ_h = Hyp mean of population of differences

$s_D^2 = \frac{\Sigma(D - \bar{D})^2}{n-1}$ Estimate of variance of population of Ds

Note: In this test we set μ_h hyp equal to zero. Thus

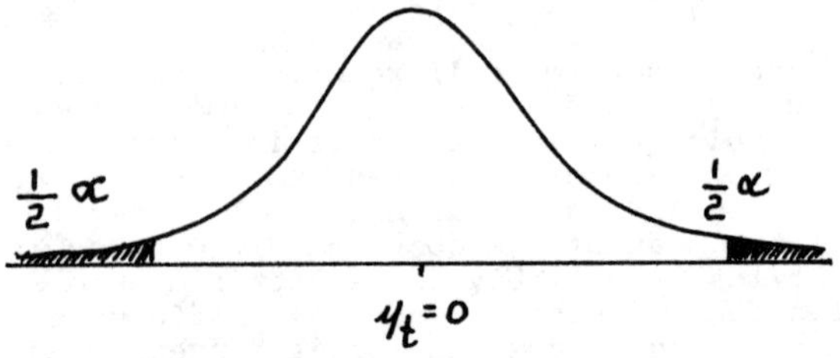

Sampling distribution of $t_{df = n-1}$ under the assumption that the hypothesis tested is true.

(5) Compute $t_{df = n-1}$ and determine whether or not it falls in a rejection region. Notice in the illustration that if we can reject the hypothesis that the mean of the population of Ds is zero, we will have reason to believe that reaction time varies with time of day.

Chapter 13

AN ENUMERATION STATISTIC: CHI SQUARE

Each of the statistical procedures that we have discussed thus far can be conceptualized as being based on the circumstances that arise when we have a set of n or more items where each item yields a measure of a given magnitude. For example, the statistic $\bar{X} = \frac{\Sigma X}{n}$ is the mean of a sample of n measures. The statistic we are about to examine χ^2 (called Chi square) applies to a quite different set of circumstances. More specifically, it deals with the situation in which rather than taking a numerical measure on each item, we merely note whether a given item falls into one or another category. For example, in a sample of 100 newly manufactured colored glass marbles we might note for a given marble whether it was perfect or flawed. Or, again, considering a set of 100 marbles, we might note for each marble whether it was black, yellow, red, or some color other than black.yellow or red.

In the case of the first kind of categories we might find that for 100 marbles 93 were perfect and 7 were flawed. In the case of the second kind of categories we might find that we had 66 black, 12 red, 15 yellow and 10 marbles of assorted other colors.

In both of the above samples we deal with a set of n = 100 items each of which falls into one of a limited number of pre specified categories. In both cases the categories are exhaustive and every item falls into one and only one category. With this arrangement, the factor of interest is the number of items in each of the several (K) categories. As we shall see, the statistic χ^2 provides us with a numerical index of the degree to which these frequencies are in agreement with the frequencies that are expected according to a given statistical hypothesis.

By now it should be clear that χ^2 is called an enumeration statistic because it deals with frequency counts. As noted above, all of the other statistics that we have considered thus far have dealt with magnitude measures rather than frequency counts.

We can gain some appreciation of the logic of χ^2 if we consider the following situation:

We have before us two barrels each containing 100,000 marbles. Here are those barrels:

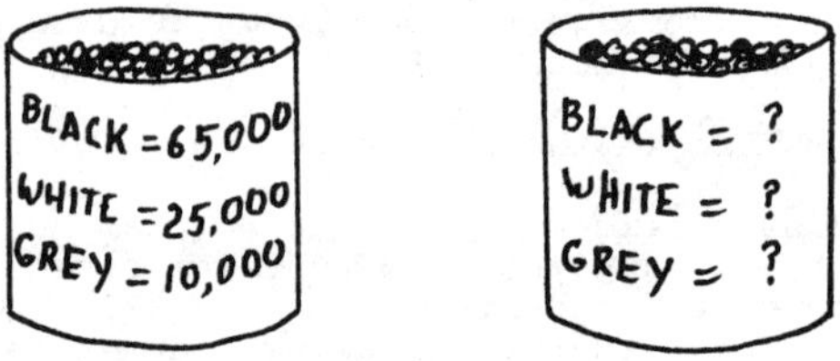

Having counted them, we know that the barrel on the left contains exactly 65,000 black, 25,000 white and 10,000 gray marbles. Our task is to draw a single sample of 100 marbles from the barrel on the right, and based on what we see in our one sample, decide whether that barrel's contents are the same as those of the barrel on the left.

We are permitted to do anything we desire with the barrel on the left, but only one sample is to be drawn from the right hand barrel. After some thought, we decide to draw a number of random samples of 100 marbles each from the barrel on the left. We reason that by doing so (and examining the results) we might learn what kinds of samples are obtained when those samples come from a population (barrel) where the proportions are as indicated on the known (left hand barrel).

Here are the frequency counts on our first 5 samples from the left hand barrel

	Sample Number				
	1	2	3	4	5
Black	70	61	72	68	64
White	25	29	23	18	23
Gray	5	10	5	14	13
Total	100	100	100	100	100

As we examine these data, we discover that the array is quite complex and cannot be readily interpreted. What seems needed is a single number (that is a single measure on each sample) that might serve as an index of the degree to which the frequencies of the various colors within a given sample conform to the frequencies that would be expected in a sample of 100 marbles from the barrel on the left.

We know that the barrel from which the samples were drawn contains 65,000 black marbles, 25,000 white marbles and 10,000 gray marbles. On this basis we can deduce that the expected frequencies of black, while and gray marbles in a sample of 100 must be 65, 25 and 10. Of course we do not anticipate that any given sample will conform to these expectations exactly, but we recognize that the numerical index we are seeking should somehow reflect the differences between the frequencies that are observed in a given sample and the above expected frequencies.

Here are those differences for sample No. 1 above:

	Observed (O)	Expected (E)	O - E
Black	70	65	(70 - 65) = 5
White	25	25	(25 - 25) = 0
Gray	5	10	(5 - 10) = -5
	$\sum O = 100$	$\sum E = 100$	$\sum(O - E) = 0$

As seen here, the differences between O and E sum to zero. Since this will always be the case when $\sum O = \sum E$, it is clear that $\sum(O - E)$ cannot serve as the index we are seeking.

One measure that could serve this purpose is $\sum(O - E)^2$.

For the data in Sample 1

$$\sum(0 - E)^2 = (70 - 65)^2 + (25 - 25)^2 + (5 - 10)^2$$
$$= (5)^2 + (0)^2 + (-5)^2$$
$$= 25 + 0 + 25$$
$$= 50$$

As we examine these figures we recognize that these calculations do in fact yield a single number that reflects the several differences between 0s and Es in Sample 1, but we can also recognize that it is deficient in an important respect. To see this deficiency we need only note that with this index ($\sum(0 - E)^2$) each deviation between 0 and E is given equal weight. Thus the squared deviation between observed and expected frequencies of black marbles $(70 - 65)^2 = 5^2 = 25$ makes the same contribution to the index (i.e., to $\sum(0 - E)^2 = 50$) as the squared difference between observed and expected frequencies of the gray marbles $(5 - 10)^2 = (-5)^2 = 25$. This does not seem appropriate because, after all, a deviation of 5 when one expects 65 is far less serious than a deviation of 5 when one expects 10. In the former case the deviation is only about 8% of the expected value, whereas in the latter case it is 50% of the expected value.

By following this line of reasoning, we come to the conclusion that the index we are seeking must somehow weight the deviations in a given category by the expected frequency in that category. An index that can accomplish this is given by the expression:

$$\chi^2 = \sum \frac{(O - E)^2}{E}$$

The next page shows the values of χ^2 for the first five samples that we obtained.

		Observed	Expected	$(O - E)^2$	$\frac{(O - E)^2}{E}$
Sample 1	Black	70	65	25	= .38
	White	25	25	0	= 0
	Grey	5	10	25	=2.5
				$\chi_1^2 = \sum \frac{(O-E)^2}{E}$	=2.88
Sample 2	Black	61	65	16	= .25
	White	29	25	16	= .64
	Grey	10	10	0	= 0
				$\chi_2^2 = \sum \frac{(O-E)^2}{E}$	= .89
Sample 3	Black	72	65	49	= .75
	White	23	25	4	= .16
	Grey	5	10	25	= .5
				$\chi_3^2 = \sum \frac{(O-E)^2}{E}$	=3.41
Sample 4	Black	68	65	9	= .13
	White	18	25	49	=1.96
	Grey	14	10	16	=1.60
				$\chi_4^2 = \sum \frac{(O-E)^2}{E}$	=3.69
Sample 5	Black	64	65	1	= .01
	White	23	25	4	= .16
	Grey	13	10	9	= .90
				$\chi_5^2 = \sum \frac{(O-E)^2}{E}$	=1.07

As we examine these data it becomes clear that the statistic $\chi^2 = \sum \frac{(O-E)^2}{E}$ is indeed the kind of measure we are seeking. When the frequencies in a given sample are close to expectations, the value of the statistic is small. When the frequencies in the sample depart from expectations, the value of the statistic is large.

Encouraged by these observations, we might continue to draw random samples (N = 100) from the left hand barrel and calculate and record the value of

$\sum\frac{(O-E)^2}{E}$ yielded by each sample.

When we have amassed several thousand values of $x^2=\sum\frac{(O-E)^2}{E}$ we would arrange the data into a frequency distribution. The histogram seen below illustrates such a distribution.

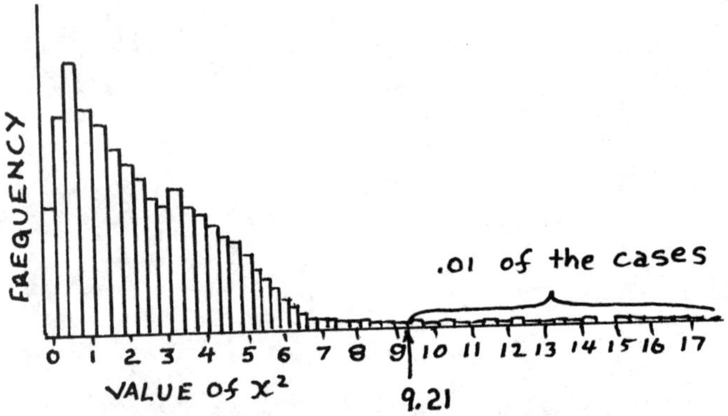

This distribution represents the values of $x^2=\sum\frac{(O-E)^2}{E}$ that are obtained when all of the samples come from a barrel in which the proportions of black to white to gray marbles is know to be 65 : 25 : 10.

An examination of the histogram reveals that under these circumstances, large values of $x^2=\sum\frac{(O-E)^2}{E}$ occur less frequently than moderate values. This reflects the fact that samples showing large deviations from the expected proportions are less likely to occur than samples with moderate deviations from the unexpected proportions.

One can also see that very small values of $\chi^2 = \sum \frac{(O-E)^2}{E}$ (which represent very close agreement between the proportions in the sample and the proportions in the barrel) are also less likely than moderate values.

Finally, as we continue to examine the histogram we discover that the value of $\chi^2 = \sum \frac{(O-E)^2}{E}$ that cuts off the upper .01 of the distribution is 9.21. Having obtained this information we would finally be prepared to draw our single sample from the barrel on the right. We would reason that if, using the Expected frequencies B = 65, W = 25, G = 10, the sample yields a value of $\chi^2 = \sum \frac{(O-E)^2}{E}$ that is equal to or larger than 9.21, we will have a basis for rejecting the hypothesis that the frequencies in the right hand barrel are identical to those of the barrel on the left.

More specifically, we deduce that there are only two reasons why the obtained value of $\chi^2 = \sum \frac{(O-E)^2}{E}$ would exceed 9.21.

(1) The proportions of black, white and gray marbles in the two barrels are identical and they just happened to have drawn a sample that was highly discrepant. Of course if this is the case, it would represent an event that only occurs one out of a hundred times.

(2) The proportions of black, white and gray marbles in the two barrels are different.

At this point we would take our random sample from the barrel on the right. If, for example, using expected frequencies of B = 65, W = 25 and G = 10, we obtained a calculated value of $\chi^2 = \sum \frac{(O-E)^2}{E} = 10.37$.

We would reject the hypothesis that the frequencies in the two barrels are identical knowing that with the decision strategy we employed, our chances of rejecting the hypothesis tested (when it was true) was only one in a hundred.

In the problem facing us, each sample contained K = 3 categories (black, white and gray) and each category contained a given number of items (0) (where 0 = the observed frequency in a given category).

Moreover, each sample contained a total of n = $\sum^{K} O$ = 100 items.

Since, in situations where χ^2 is calculated, the $\sum^{K} E$ must also equal n. The maximum number of degrees of freedom in χ^2 is K - 1. This is because in calculating the expected values for each category, the value for the last category is always determined. For any set of data, it must be that value which permits $\sum^{K} E$ to equal n. This, in turn, means that in calculating the values of each of the K $\frac{(O-E)^2}{E}$s that make up the χ^2, the final $\frac{(O-E)^2}{E}$ is completely determined.

In short, in the application of χ^2 that we have been discussing only K - 1 of its elements are free to vary and hence χ^2 has K - 1 df.

In general, the sampling distribution of χ^2 is determined by its degrees of freedom. When dfs are small the sampling distribution of χ^2 is highly skewed (as was the case when we obtained an empirical sampling distribution of χ^2 with df = 2). As its df increase (this happens when the number of categories (K) is increased) the mean of the sampling distribution of χ^2 increases, its variance decreases, and the distribution begins to approximate a normal function. Table VI shows the values of χ^2 for each of a number of different dfs that cut off various percentages of the sampling distribution that obtains when the expected frequency for each category in the sample is based on the relative frequency of items in the corresponding category of the population.

It was stated above that as df increases the mean of the sampling distribution of χ^2 increases. In point of fact, when the hypothesis tested is true, the mean of the sampling distribution of χ^2 _is its df_. As seen in the example we started with, the mean of the empirical sampling distribution of χ^2 that we obtained was 2. In this case there were K = 3 categories and hence K - 1 = 2 df.

Earlier we studied the sampling distribution of the statistic $\frac{s_x^2}{\sigma_x^2}$. That statistic, like χ^2, also has a sampling distribution that is determined by its df, but in the case of $\frac{s_x^2}{\sigma_x^2}$ the mean of the sampling distribution (when the samples came from a population with variance equal to σ_x^2) was always one. It may not be obvious at first, but the sampling distribution of $\frac{s_x^2}{\sigma_x^2}$ can be converted into a sampling distribution of χ^2 by multiplying every $\frac{s_x^2}{\sigma_x^2}$ in the distribution by its df. You can check this point for yourself by comparing the tabled values in Table VI to the tabled values in Table II. As can be seen, the values in Table VI are the values in Table II multiplied by the df.

Chapter 14

ANALYSIS OF VARIANCE: SINGLE VARIABLE OF CLASSIFICATION

Up to now we have examined a variety of ways to decide whether or not an observed difference between two means is statistically reliable. At this point we will examine a technique for deciding whether or not the observed differences among a set of K means is statistically reliable. The technique is called analysis of variance because it involves an analysis of the variability among the measures. It is important, however, to recognize that the analysis of variance is designed to enable us to make decisions about a set of observed means. It is not (as some unwitting students have been known to suggest) a procedure to test for differences among variances.

To illustrate the logic of analysis of variance, we will consider the situation in which we have K samples with n items in each. Here is a set of K = 4 such samples with n = 7 items in each.

Sample 1

$$\left.\begin{matrix} X_1 \\ X_2 \\ \vdots \\ X_7 \end{matrix}\right\} \quad \bar{X}_1 = \frac{\sum X}{n} \qquad s_X^2 = \frac{\sum (X - \bar{X}_1)^2}{n-1}$$

Sample 2

$$\left.\begin{matrix} X_1 \\ X_2 \\ \vdots \\ X_7 \end{matrix}\right\} \quad \bar{X}_2 = \frac{\sum X}{n} \qquad s_X^2 = \frac{\sum (X - \bar{X}_2)^2}{n-1}$$

Sample 3

$$\left.\begin{matrix} X_1 \\ X_2 \\ \vdots \\ X_7 \end{matrix}\right\} \quad \bar{X}_3 = \frac{\sum X}{n} \qquad s_X^2 = \frac{\sum (X - \bar{X}_3)^2}{n-1}$$

Sample 4

$$\left.\begin{matrix} X_1 \\ X_2 \\ \vdots \\ X_7 \end{matrix}\right\} \quad \bar{X}_4 = \frac{\sum X}{n} \qquad s_X^2 = \frac{\sum (X - \bar{X}_4)^2}{n-1}$$

Our task is to decide whether or not the observed differences among the four means of the samples are statistically reliable. In other words, we must determine whether we have any basis to reject the hypothesis that the samples come from populations with the same means. Note, however, that the analysis assumes homogeneity of variance, that is, the procedure assumes that the samples come from populations with the same variance. This fact, in combination with the nature of the statistical hypothesis we are testing (namely that the samples come from populations with the same means) enable us to conceptualize the analysis of variance as a test of the hypothesis that the K samples come from a single population with a variance equal to σ_X^2.

If this hypothesis is true, we can use the data in our K sample to make two independent estimates of the variance (σ_X^2) of the population from which the samples are presumably drawn. One of these estimates is based upon the observed variability within our samples. The other is based upon the observed variability among the means of our samples.

Consider first the estimate of σ_X^2 that is based upon the observed variability within our samples.

For any one sample we know that: $s_X^2 = \dfrac{\sum^{n} (X - \bar{X})^2}{n - 1}$

is an unbiased estimate of σ_X^2. i.e., $s_X^2 \rightarrow \sigma_X^2$

Accordingly, the average of the several s_X^2s from the samples is also an unbiased estimate of σ_X^2. The symbol s_p^2 will represent this average where

$$s_p^2 = \frac{\sum^{K} s_X^2}{K} = \frac{1}{K}\left[\frac{\sum^{n}(X - \bar{X}_1)^2}{n-1} + \frac{\sum^{n}(X - \bar{X}_2)^2}{n-1} + \cdots + \frac{\sum^{n}(X - \bar{X}_K)^2}{n-1}\right]$$

$$s_p^2 = \frac{\sum^{K}\sum^{n}(X - \bar{X})^2}{K(n-1)}$$

In short, s_p^2 is an estimate of σ_X^2 that is based upon the average of the s_X^2 s within our samples.

Our second estimate of σ_X^2 is based upon the observed variability among the K means (tempered with a knowledge of what happens when one forms a sampling distribution of means).

We know (from the central limit theorem) that if we form a sampling distribution of means based upon samples of size n, the following will be true:

$\sigma_{\bar{X}}^2 = \dfrac{\sigma_x^2}{n}$ i.e., the variance of the sampling distribution of means ($\sigma_{\bar{X}}^2$) will be equal to the variance of the population from which the samples are drawn (σ_X^2) divided by the size of the samples (n).

When conceptualized in these terms, the means of the K samples (in our analysis of variance problem) constitute a sample (of size K) from the sampling distribution of means.

Accordingly:

$$s_{\bar{X}}^2 = \frac{\sum^{K} (\bar{X} - \bar{\bar{X}})^2}{K - 1}$$

Note: $\bar{X}$ is the mean of one of the K samples and $\bar{\bar{X}}$ is the mean of the K means.

is an unbiased estimate of $\sigma_{\bar{X}}^2$

i.e., $s_{\bar{X}}^2 \longrightarrow \sigma_{\bar{X}}^2$

Since, however, $\sigma_{\bar{X}}^2 = \dfrac{\sigma_X^2}{n}$ and $n\sigma_{\bar{X}}^2 = \sigma_X^2$

$n s_{\bar{X}}^2$ must be an unbiased estimate of σ_X^2

i.e., $n\, s_{\bar{X}}^2 \longrightarrow \sigma_{\bar{X}}^2$

In short, $n\, s_{\bar{X}}^2$ is an estimate of σ_X^2 that is based upon the variations among the means of our K samples.

Consider now what would happen if we engage in the following exercise:

We start with a population with a given mean and variance and randomly select K samples of size n.

We then compute $n s_{\bar{X}}^2$ and s_p^2 and form the ratio $\dfrac{n s_{\bar{X}}^2}{s_p^2}$

Our first set of K samples would yield some particular value of the statistic. For example: .93

Now we go back to the population and draw a second set of K samples. We again compute $\dfrac{n s_{\bar{X}}^2}{s_p^2}$ and this time get a different value. For example: 1.13.

If we repeated this procedure indefinitely, we would obtain a sampling distribution of the statistic

$$\frac{n s_{\bar{X}}^2}{s_p^2}$$

Since $n s_{\bar{X}}^2 \rightarrow \sigma_X^2$ and $s_p^2 \rightarrow \sigma_X^2$,

the ratio $\dfrac{n s_{\bar{X}}^2}{s_p^2}$ should tend to approximate 1 (one).

In other words, the mean of the sampling distribution will be one. Sometimes the statistic will have a value greater than 1 and sometimes its value will be les than 1, but on the average, it will approximate 1.

In general, the shape of the sampling distribution will depend upon the number of samples that contribute to the statistic and upon the size of the samples (i.e., for each different combination of K and n , we will get a different sampling distribution of the statistic $\dfrac{n s_{\bar{X}}^2}{s_p^2}$).

Here is the sampling distribution that would be obtained if, for example, we repeatedly drew 5 samples of 11 items each from a given normal population and computed the statistic $\frac{n s_{\bar{x}}^2}{s_p^2}$ on each set of 5 samples.

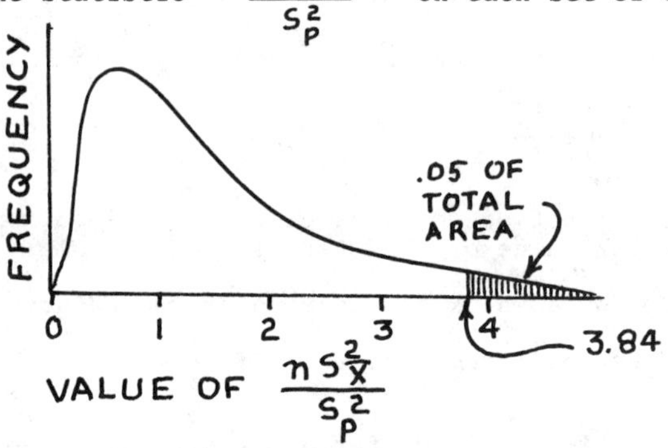

This distribution would have a mean of one. The shaded portion shows those extremely large values that would occur only 5% of the time. As seen here, the value that cuts off the upper 5% of the distribution is 3.84.

Now let us return to our original problem. We have one set of K samples with n items in each. We must decide whether or not the observed differences between the means is reliable. That is, we must decide whether or not to reject the statistical hypothesis which asserts that the samples came from the same population.

Since the above sampling distribution is the one that occurs when the K samples do, in fact, come from the same population, our decision process is quite straightforward. All we need to do is to compute

$\frac{n s_{\bar{x}}^2}{s_p^2}$ on our one set of K samples and observe

whether or not the value which is obtained falls in the rejection region of the above sampling distribution.

If it does fall in the rejection region, then our one set of K samples has led to an event which would only occur 5% of the time when the samples are drawn from the same population (i.e., when the statistical hypothesis is true). We, therefore, reject the statistical hypothesis at the = .05 level of confidence and assume that the observed difference among our means are not a result of sampling variations.

If our obtained statistic does not fall in the rejection region, we have no basis to reject the null hypothesis, i.e., to assume that the observed differences among our means represent anything other than random variations that occur in the course of taking samples.

Hopefully, by now it is obvious to you that when the samples come from the same population, the statistic $\frac{n s_{\bar{X}}^2}{s_p^2}$ is in fact a ratio of two estimates of a population variance (σ_X^2) and just as the statistic $\frac{s_{X_1}^2}{s_{X_2}^2}$ is distributed as F with (n_1 - 1) df for the numerator and (n_2 - 1) df for the denominator the statistic $\frac{n s_{\bar{X}}^2}{s_p^2}$ will be distributed as F with degrees of freedom equal to (K - 1) for the numerator and K(n - 1) for the denominator. Accordingly, we can use the F tables to find the appropriate critical ratios. For a test with α = .05 we would use the table in Appendix V. For a test with α = .01 we would use the table in Appendix VI.

The great versatility of the Analysis of Variance technique has led to the standardization (and simplification) of procedures for computing and reporting Fs.

The next page shows two versions of the standard format for reporting an analysis of variance.

The top version incorporates what some statisticians describe as the conceptual formula for the analysis of variance. These are the elements that when properly combined make up the expression

$$F_{df = K-1,\ K(n-1)} = \frac{n s_{\bar{X}}^2}{s_p^2}$$

The bottom version incorporates the computing formula for the identical analysis. It is important to recognize that despite their seemingly different structures the conceptual and computing versions of the two formulas are algebraically equivalent and hence for a given set of data they will always yield identical results.

	SUM OF SQUARES	DF	MEAN SQUARE	F
BETWEEN	$n\sum^{K}(\bar{X} - \bar{\bar{X}})^2$	$K-1$	$\frac{\text{SS BET}}{\text{DF BET}}$	$\frac{\text{MS BET}}{\text{MS WIT}}$
WITHIN	$\sum^{K}\sum^{n}(X - \bar{X})^2$	$K(n-1)$	$\frac{\text{SS WIT}}{\text{DF WIT}}$	
TOTAL	$\sum^{K}\sum^{n}(X - \bar{\bar{X}})^2$	$Kn-1$		

	SUM OF SQUARES	DF	MEAN SQUARE	F
BETWEEN	$\sum^{K}\frac{(\sum^{n} X)^2}{n} - \frac{(\sum^{K}\sum^{n} X)^2}{Kn}$	$K-1$	$\frac{\text{SS BET}}{\text{DF BET}}$	$\frac{\text{MS BET}}{\text{MS WIT}}$
WITHIN	$\sum^{K}\sum^{n} X^2 - \sum^{K}\frac{(\sum^{n} X)^2}{n}$	$K(n-1)$	$\frac{\text{SS WIT}}{\text{DF WIT}}$	
TOTAL	$\sum^{K}\sum^{n} X^2 - \frac{(\sum^{K}\sum^{n} X)^2}{Kn}$	$Kn-1$		

Here is a proof that

$$n\sum^{K}(\bar{X} - \bar{\bar{X}})^2 = \sum^{K}\frac{(\sum^{n} X)^2}{n} - \frac{(\sum^{K}\sum^{n} X)^2}{nK}$$

Earlier it was shown that

$$\sum(X - \bar{X})^2 = \sum X^2 - \frac{(\sum X)^2}{n}$$ (See page 145.)

This is the computational formula for the sum of n squared deviations from the mean.

In the present context we must deal with n times the sum of the squared deviations of K means from the mean of the K means.

i.e., we must deal with $n\sum^{K}(\bar{X} - \bar{\bar{X}})^2$

We now make use of the computational formula to rewrite the above expression as:

$$n\sum^{K}(\bar{X} - \bar{\bar{X}})^2 = n\left[\sum^{K}\bar{X}^2 - \frac{(\sum^{K}\bar{X})^2}{K}\right]$$

We now substitute $\frac{\sum^{n} X}{n}$ for $\bar{X}$ and obtain

$$n\left[\sum^{K}\left(\frac{\sum^{n} X}{n}\right)^2 - \frac{\left(\sum^{K}\frac{\sum^{n} X}{n}\right)^2}{K}\right]$$

It will be recalled (or learned right now) that $\sum\left(\frac{X}{C}\right)^2$ can be written as $\frac{1}{C}\sum\frac{X^2}{C}$ and that $\left(\sum\frac{X}{C}\right)^2$

can be written as $\frac{1}{c^2}\left[\sum X\right]^2$

When we move n through the summation sign in this fashion, our formula becomes

$$n\left[\frac{1}{n}\sum^{K}\frac{\left(\sum^{n} X\right)^2}{n} - \frac{1}{n^2}\frac{\left(\sum^{K}\sum^{n} X\right)^2}{K}\right]$$

Cancelling n where possible, we obtain:

$$\sum^{K}\frac{\left(\sum^{n} X\right)^2}{n} - \frac{\left(\sum^{K}\sum^{n} X\right)^2}{nK}$$

which is what we set out to demonstrate, namely that

$$n\sum^{K}(\bar{X}-\bar{\bar{X}})^2 = \sum^{K}\frac{\left(\sum^{n} X\right)^2}{n} - \frac{\left(\sum^{K}\sum^{n} X\right)^2}{nK}$$

In short, we have shown that the between sum of squares is equal to n times the sum of the squared deviations of the K means from the mean of the K means. Once this has been shown it is a simple step to see that the between sum of squares divided by its degrees of freedom is numerically equal to $n\,s_{\bar{X}}^2$

The proof that the sum of squares for within divided by its degrees of freedom is equal to s_p^2 can be approached in the same fashion.

The next page shows that proof:

$$s_p^2 = \frac{\sum^{K} s_X^2}{K} = \frac{s_{X_1}^2 + s_{X_2}^2 + \cdots s_{X_K}^2}{K}$$

$$= \frac{1}{K}\left[\frac{\sum(X-\bar{X}_1)^2}{n-1} + \frac{\sum(X-\bar{X}_2)^2}{n-1} + \cdots + \frac{\sum(X-\bar{X}_K)^2}{n-1}\right]$$

$$= \frac{1}{K(n-1)}\left[\sum^{K}\sum^{n}(X-\bar{X})^2\right]$$

$$= \frac{1}{K(n-1)}\left[\sum^{K}\left(\sum^{n}X^2 - \frac{\left[\sum^{n}X\right]^2}{n}\right)\right]$$

$$= \frac{\sum^{K}\sum^{n}X^2 - \sum^{K}\frac{\left(\sum^{n}X\right)^2}{n}}{K(n-1)}$$

It may not be intuitively obvious at first, but we can also show that

$$\sum^{K}\sum^{n}(X-\bar{\bar{X}})^2 = \sum^{K}\sum^{n}X^2 - \frac{\left(\sum^{K}\sum^{n}X\right)^2}{nK}$$

That is, the total SS in the computing formula on page 239 is numerically equal to the sum of squared deviation of each of the Kn Xs from the grand mean of those Kn Xs. To prove this statement, we again make use of the computing formula (see page 145) which says that for a set of n deviations from the mean of the set

$$\sum^{n}(x-\bar{x})^2 = \sum^{n} x^2 - \frac{(\sum^{n} x)^2}{n}$$

In the present case we are dealing with a set of Kn deviations from the mean of the set

Thus $$\sum^{K}\sum^{n}(x-\bar{\bar{x}})^2 = \sum^{K}\sum^{n} x^2 - \frac{(\sum^{K}\sum^{n} x)^2}{nK}$$

If we carefully examine the analysis of variance table on page 239, we discover the interesting fact that whenever we have K sets of n items the total sum of squares is equal to the within sum of squares plus the between sum of squares. To prove this statement we need only write it in algebraic form. Here is that expression:

$$\underbrace{\sum^{K}\sum^{n}(x-\bar{\bar{x}})^2}_{\text{TOTAL SS}} = \underbrace{\sum^{K}\sum^{n}(x-\bar{x})^2}_{\text{WITHIN SS}} + \underbrace{n\sum^{K}(\bar{x}-\bar{\bar{x}})^2}_{\text{BETWEEN SS}}$$

TOTAL SS = WITHIN SS + BETWEEN SS

$$\overbrace{\sum^{K}\sum^{n}x^2 - \frac{(\sum^{K}\sum^{n}x)^2}{Kn}} = \overbrace{\sum^{K}\sum^{n}x^2 - \sum^{K}\frac{(\sum^{n}x)^2}{n}} + \overbrace{\sum^{K}\frac{(\sum^{n}x)^2}{n} - \frac{(\sum^{K}\sum^{n}x)^2}{Kn}}$$

Notice that in the right hand side of the computing formula the term $\sum^{K} \frac{(\sum^{n} x)^2}{n}$ is both added and subtracted and hence it drops out.

With some thought it should become clear that we have here a variety of different ways of saying the same things. We can summarize these important statements in one more way by asserting that whenever we have a total of Kn observations (items) that are arranged in K sets of n items the sum of the squared deviations of those Kn items from their mean can be partitioned into two components.

1) A component that is equal to the sum of the squared deviation of each item from the mean of the sub set in which it appears, i.e.,

$$\sum^{K}\sum^{n}(x-\bar{x})^2$$

2) A component that is equal to n times the sum of the squared deviations of the means of the sub sets from the grand mean, i.e.,

$$n\sum^{K}(\bar{x}-\bar{\bar{x}})^2$$

A NUMERICAL EXAMPLE OF ONE WAY ANALYSIS OF VARIANCE

We can gain some appreciation of these formulas and the relationships among them if we consider a numerical example. Suppose an experimenter wishes to examine the manner in which the rat's tendency to perform a food reinforced lever press is influenced by the animal's motivational state. To study this issue each of 18 rats is trained to press a lever using food as reinforcement. When all of the animals have learned the response to a given criterion, the subjects are fed to satiety, and then are randomly assigned to one of three deprivation conditions: 10 hours, 20 hours, 30 hours. After allowing the animals to go without food for the appropriate intervals, they are tested again in the lever press apparatus.

Here are the numbers of presses recorded from each of the six rats in a given group during this test.

	No. of Lever Presses					
Group I (10 hrs deprivation):	3,	1,	5,	4,	3,	2
Group II (20 hrs deprivation):	10,	7,	8,	7,	6,	4
Group III (30 hrs deprivation):	7,	9,	8,	10,	8,	6

The first step in analyzing these data is to try to depict them in a fashion that will reveal their general trend. One way to do this is to calculate the mean of each group and to plot the means in graphic form.

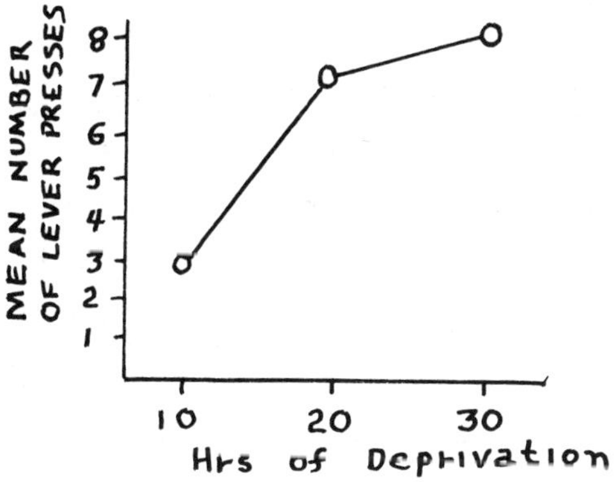

As seen here, the animals in the 30 hour group pressed more, on the average, than the animals in the 20 hour group, and those animals pressed more than the 10 hour deprivation animals. Clearly, the data are consistent with the proposition that the rats' tendency to exhibit food reinforced behavior is an increasing function of the number of prior hours of deprivation. Of course, there is another way to account for the observed differences between the means of the three groups. It is possible that the experimental operation (hours of deprivation) was completely without effect and the observed differences between means is a random event than can be attributed to the sampling variations that occurred during the formation of the three groups.

As we have indicated right along, the analysis of variance helps us to decide between these possibilities. It does so by enabling us to determine if we can reject the hypothesis that the observed variation among the means can be attributed to chance. Another way to say this is that the analysis of variance enables us to determine if the observed differences among the three means is one of those rare events that would only occur say one percent of the time if we repeatedly took three samples of six rats each and tested all three groups under the same level of deprivation.

The calculations on the next several pages illustrate the details of this analysis and they reveal the several numerical equivalences that characterize the computations in the analysis.

The calculations on page 248 approach the issue from the conceptual framework that was employed at the beginning of this chapter (when we introduced the analysis of variance). The calculations on the following two pages show the same analysis using the computational procedure that is generally used in practice. As can be seen, both methods yield exactly the same value of F, i.e., $F_{df=2,15} = 15.75$

If we look into Table IV we discover that for

$F_{df=2,15}$ with $\alpha = .01$ the critical value is

6.36. Since 15.75 > 6.36 we reject the hypothesis that

the observed differences are attributable to chance variations in sampling. That is, we reject the idea that the observed effect is one of those chance occurrences that are expected to appear 5% of the time when our treatment has no effect. Accordingly, we assume that the observed differences among the means was due to the differences in hours of deprivation in the three groups.

Admittedly, the material we have covered is complex and it offers many opportunities for conceptual pitfalls, but if one carefully studies the calculations on the next few pages, the inherent logic of the procedure will become increasingly clear.

3
1
5
4
3
2

$$\bar{X}_1 = \frac{18}{6} = 3$$

$$S^2_{X_1} = \frac{10}{5} = 2$$

SAMPLE 1

10
7
8
7
6
4

$$\bar{X}_2 = \frac{42}{6} = 7$$

$$S^2_{X_2} = \frac{20}{5} = 4$$

SAMPLE 2

7
9
8
10
8
6

$$\bar{X}_3 = \frac{48}{6} = 8$$

$$S^2_{X_3} = \frac{10}{5} = 2$$

SAMPLE 3

$$\bar{\bar{X}} = \frac{\sum^{K} \bar{X}}{K} = \frac{3+7+8}{3} = \frac{18}{3} = 6$$

$$S^2_{\bar{X}} = \frac{\sum^{K} (\bar{X} - \bar{\bar{X}})^2}{K-1}$$

$$\sum^{K} (\bar{X} - \bar{\bar{X}})^2 = (3-6)^2 + (7-6)^2 + (8-6)^2 = 14$$

$$n S^2_{\bar{X}} = 6 \frac{14}{2} = 42$$

$$S^2_p = \frac{S^2_{X_1} + S^2_{X_2} + S^2_{X_3}}{3} = \frac{2+4+2}{3} = 2.6\dot{6}$$

$$F_{df=2,15} = \frac{42}{2.6\dot{6}} = 15.75$$

	X	X^2	X	X^2	X	X^2
	3	9	10	100	7	49
	1	1	7	49	9	81
	5	25	8	64	8	64
	4	16	7	49	10	100
	3	9	6	36	8	64
	2	4	4	16	6	36
$\sum^{n}$	18	64	42	314	48	394

$\sum^{K}\sum^{n} X^2 = 64 + 314 + 394 = 772$	
$\sum^{K}(\sum^{n} X)^2 = 18^2 + 42^2 + 48^2 = 4392$	
$\sum^{K}\sum^{n} X = 18 + 42 + 48 = 108$	$n=6$ $K=3$ $nK=18$
$(\sum^{K}\sum^{n} X)^2 = 108^2 = 11664$	$K-1 = 3-1 = 2$
$\sum^{K}\frac{(\sum^{n} X)^2}{n} = \frac{4392}{6} = 732$	$K(n-1) = 3(6-1) = 15$
$\frac{(\sum^{K}\sum^{n} X)^2}{nK} = \frac{11664}{18} = 648$	$Kn-1 = 18-1 = 17$

	SUM OF SQUARES	DF	MEAN SQUARE	F
BETWEEN	$\sum^{K} \frac{(\sum^{n} X)^2}{n} - \frac{(\sum^{K}\sum^{n} X)^2}{Kn}$	$K-1$	$\frac{SS\ BET}{DF\ BET}$	$\frac{MS\ BET}{MS\ WIT}$
WITHIN	$\sum^{K}\sum^{n} X^2 - \sum^{K} \frac{(\sum^{n} X)^2}{n}$	$K(n-1)$	$\frac{SS\ WIT}{DF\ WIT}$	
TOTAL	$\sum^{K}\sum^{n} X^2 - \frac{(\sum^{K}\sum^{n} X)^2}{Kn}$	$Kn-1$		

	SUM OF SQUARES	DF	MEAN SQUARE	F
BETWEEN	732 − 648 = 84	2	$\frac{84}{2}$ = 42	$\frac{42}{2.6\dot{6}}$ = 15.75
WITHIN	772 − 732 = 40	15	$\frac{40}{15}$ = $2.6\dot{6}$	
TOTAL	772 − 648 = 124	17		

An Additional Comment

When the topic of analysis of variance was introduced it was asserted that its procedures apply to the situation in which we have K samples of n items each and we must decide whether or not the observed differences among the means of the samples can be attributed to the different treatments (or conditions) that prevailed for each sample.

As noted at that time, the analysis we conduct to resolve this issue can be described as a test of the statistical hypothesis that the several samples came from a single population. Another (and in certain respects more meaningful) way to describe such an analysis is to assert that it is a test of the hypothesis that the several treatments (or conditions) have no differential effects. In other words, we phrase our hypothesis in such a way that if our analysis yields a significant value of F we can conclude that the treatments (or conditions of the study have made our sample means more variable than they would have otherwise been.

To see how this could come about, consider the situation that might arise if we were to conduct an experiment in which we select 3 random samples of n = 4 subjects in each and we apply treatment 1 to all members in sample 1, we aply treatment 2 to all members in sample 2 and we apply treatment 3 to all members in sample 3.

The top of the next page illustrates the kind of data that would result if our treatments had no effects whatsoever. More specifically, it shows the values of X that might be observed when a zero treatment effect is added to each original X.

Original X	9	10	8	10	$\bar{X}_1 = 9.25$
Treatment 1	0	0	0	0	
Observed X	9	10	8	10	$s^2_{X_1} = .92$
Original X	10	10	9	11	$\bar{X}_2 = 10$
Treatment 2	0	0	0	0	
Observed X	10	10	9	11	$s^2_{X_2} = 1.0$
Original X	8	9	11	11	$\bar{X}_3 = 9.25$
Treatment 3	0	0	0	0	
Observed X	8	9	11	11	$s^2_{X_3} = 1.58$

$s^2_p = 1.17$ $\quad n s^2_{\bar{X}} = 1.13$ $\quad F_{df=2,9} = .97$

If, as seen here, our 3 treatments are without effect (i.e., they only add zero to each item), the observed differences among the means would reflect only the chance factors that occur when one draws random samples and accordingly our chances of obtaining a significant value of F is precisely the level we set for α (either .05 or .01).

Shown, below are the same three original samples as above, but this time they have been modified to show how they would appear if, rather than adding 0, treatment 1 added 1 point to every item in sample 1, treatment 2 added 3 points to every item in sample 2 and treatment 3 added 4 points to every item in sample 3.

Original X	9	10	8	10	$\bar{X}_1 = 10.25$
Treatment 1	1	1	1	1	
Observed X	10	11	9	11	$s^2_{X_1} = .92$
Original X	10	10	9	11	$\bar{X}_2 = 13$
Treatment 2	3	3	3	3	
Oberved X	13	13	12	14	$s^2_{X_2} = 1.0$
Original X	8	9	11	11	$\bar{X}_3 = 13.25$
Treatment 3	4	4	4	4	
Observed X	12	13	15	15	$s^2_{X_3} = 1.58$

$s^2_p = 1.17$ $\quad n s^2_{\bar{X}} = 11.08$ $\quad F_{df=2,9} = 9.47$

As can be seen, the s^2_X calculated on each sample remains unchanged, hence s^2_p (i.e., MS within) is unchanged, but $n\,s_{\bar{X}}^2$ (i.e., MS Between) calculated on the data has been inflated by an amount that is determined by the differences among the treatment effects.

When conceptualized in this way a large (significant) F in the analysis of variance would lead us to reject the hypothesis that the treatment effects were 0 because it would be more likely that the treatments had (as in the above example) driven our $\bar{X}$s (but not our Xs within samples) farther apart than they would have otherwise been.

Chapter 15

TWO-WAY ANALYSIS OF VARIANCE

The two-way analysis of variance employs a logic that is similar to that of the one-way analysis of variance. It differs from the one-way design in that the two-way design applies to those situations in which each of our measures has been obtained on a subject (or item) that has been subjected to some combination of two kinds of treatments. (In the one-way design each subject receives only one kind or level of treatment.) For example, each of 60 subjects is is tested for accuracy of throwing darts at a target under one of the 12 possible combinations of three levels of room illumination and four room temperatures. If we record the total score that each subject obtains, and if for the test we had randomly assigned 5 subjects to each combination of testing conditions, the data from the study could be arranged in the following 3 x 4 array:

Each X is the total score obtained by a given subject.

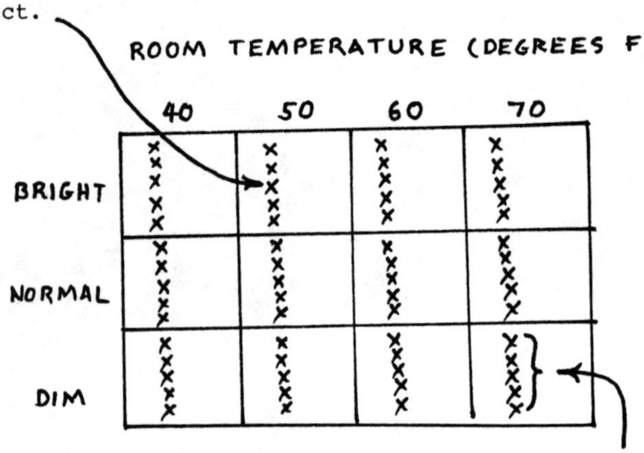

These 5 subjects were tested in a room that was dimly illuminated with the temperature set at 70 degrees.

An experiment of this sort is called a complete factorial design with replication. The design is a complete factorial because all possible combinations of the 3 levels of illumination and 4 levels of temperature appear in it. The design is said to incorporate replications because there are more than one subject per combination (i.e., the design is replicated 5 times).

The analysis of data such as these proceeds along the lines employed in the one-way analysis discussed earlier. In both cases we ask what parts of the observed variability in our data can be attributed to random sampling and what parts can be attributed to effects of the treatments.

To carry out this analysis it will be convenient to use the symbol C to refer to the number of columns (temperatures, of which there are 4). The symbol R will refer to the number of rows (levels of illumination of which there are 3). And the symbol n will refer to the number of measures under a given combination of treatments (there are 5 measures per cell).

Using these symbols we see that there are a total

of nRC = (5)(4)(3) = 60 measures (i.e., subjects). We also see that if each measure is conceived to be an X

the quantity $\sum^{C}\sum^{R}\sum^{n} X$ will be the total of all

of the measures and the quantity $\bar{\bar{X}} = \frac{\sum^{C}\sum^{R}\sum^{n} X}{CRn}$

will be the grand mean of those 60 measures. Our strategy in analyzing these data will be to assess the degree to which the variability of the 60 measures around the grand mean is organized according to the various kinds of treatments employed. Thus we will want to determine what part of the total variability in the data can be attributed to the effects of room temperature, what part can be attributed to the effects of illumination and what part can be attributed to other factors such as the random variations that inevitably occur when we take random samples.

As we have suggested earlier, this is the same strategy that we employed in the one-way analysis but in the present instance we are concerned with dissecting out the effects of two (rather than one) kind of treatment. In general the effects of a given treatment variable are reflected in the differences among the means of subjects that are tested under the several levels of that treatment.

In the example we are studying 20 subjects are tested in bright illumination, 20 subjects are tested in normal illumination and 20 are tested in dim light. If the treatment variable: room illumination, influences our measure, we would expect that the mean of the subjects exposed to one level of the treatment (for example, bright illumination) would differ from the mean of the subjects exposed to other levels of the treatment (normal or dim illumination).

Here is a convenient way to visualize these and the several other means that we can calculate with a design of this sort.

We begin with a set of items arrange in R rows and C columns with n items per cell.

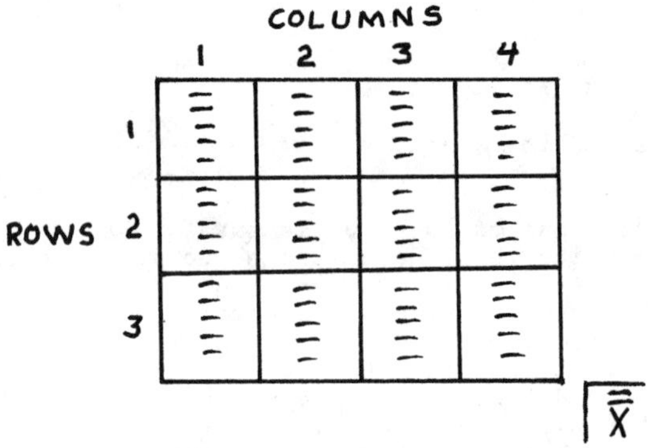

We note first that the entire array has a mean

which we can represent by the symbol $\bar{\bar{X}}$. As seen on the preceding page, $\bar{\bar{X}}$ is based on nCR items.

Next we note that each cell has a mean which we can represent by the symbol .

As seen here there are RC cell means and each is based on n items.

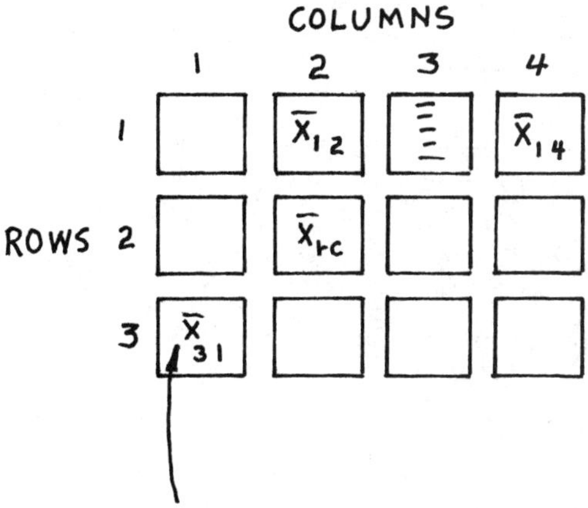

This is the mean of the cell that occupies the third row of the first column.

The next page shows the data used to calculate the means of the several (R) rows. In that array, the symbol $\bar{X}_r$ represents the mean of the rth row. It is based on the nC items in the row.

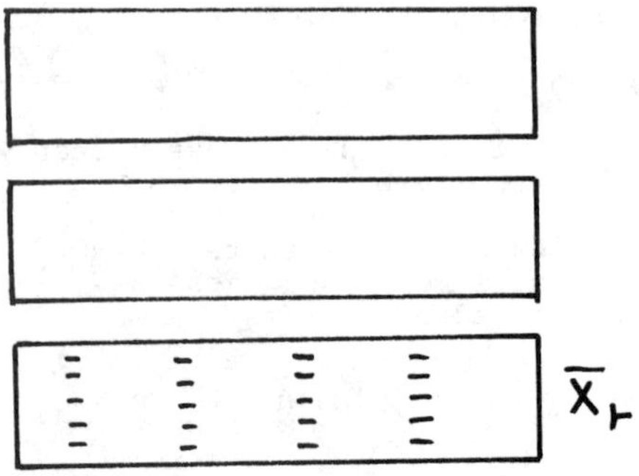

Note that the value of $\bar{X}_r$ for a given row will be the same whether we calculate it as the mean of the nC items in that row or as the mean of the C values of $\bar{X}_{rc}$ within that row.

The next figure shows the data used to calculate the means of the several (C) columns. With this usage the symbol $\bar{X}_c$ represents the mean of the Cth column. It is based on the nR items in the column.

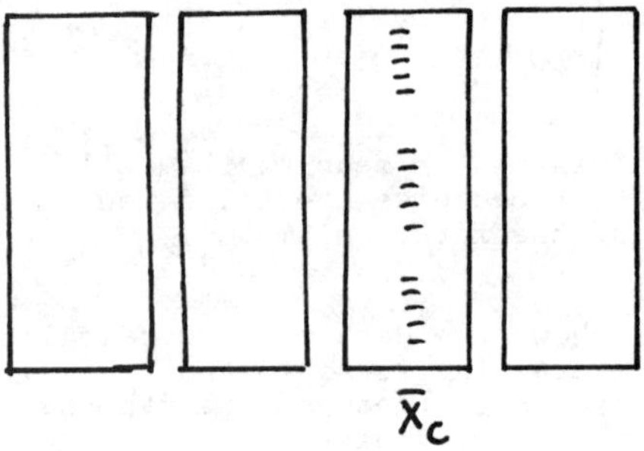

Note also that the value of $\bar{X}_c$ for a given column will be the same whether we calculate it as the mean of the nR items in that column or as the mean of the R values of $\bar{X}_{rc}$ within that column.

Though it is perhaps not immediately obvious, there is another important set of means that we can calculate in a factorial design. They can be described as cell means that have been corrected for the observed row and column effects.

To envision these corrected cell means we first need to recognize that in an experiment of this sort the effect of a given row treatment will be reflected in the observed difference between the mean for that row and the grand mean. The figure shown below illustrates the comparisons involved in assessing those differences.

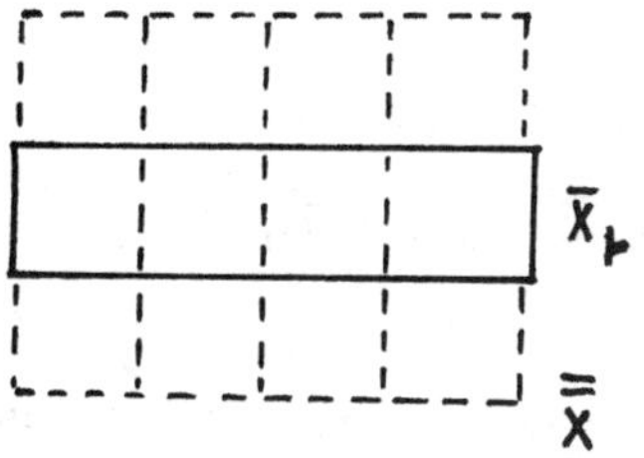

The effect of the treatment in the rth row is reflected in the quantity $(\bar{X}_r - \bar{\bar{X}})$

To see why we need to recall that in a factorial experiment (as in a one-way analysis) we conceive of a given row treatment as adding a constant amount to every item in the appropriate row. This means that if the several row treatments are producing differential effects (i.e., the treatment applied to the items in row 1 adds an amount that is different from amounts added by the treatment applied to the items in rows 2 and 3) the means of the rows will be different from each other and for a given row treatment the quantity $(\bar{X}_r - \bar{\bar{X}})$ will reflect those differences. Similarly, the effect of a given column treatment will be reflected in the observed difference between the mean of that column and the grand mean.

Here are the comparisons involved in assessing these column effects.

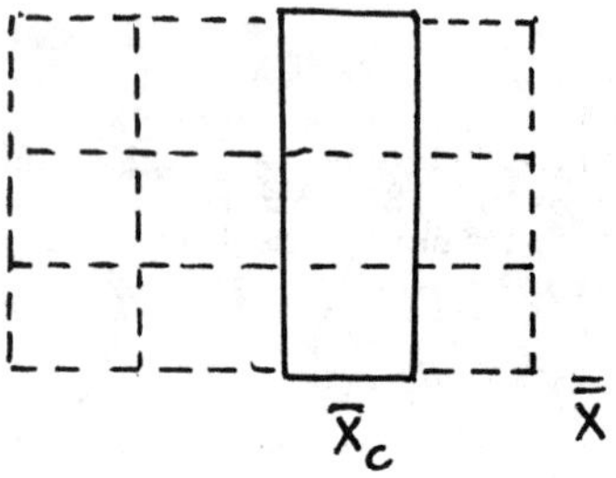

The effect of the treatment in the cth column is reflected in the quantity $(\bar{X}_c - \bar{\bar{X}})$

Of course, in a factorial experiment each item in a given cell is subjected to both a row and a column treatment and this raises the question of how the two treatments are combining. More specifically, it raises the question of whether the two kinds of treatments have effects that are independent of each other. Are, for example, the effects of the cth column treatment the same for the items in each of the R cells in the cth column?

As seen above, the quantity $(\bar{X}_c - \bar{\bar{X}})$ is an index of the overall effect of the treatment applied to the items in the cth column, but, as also seen above, some of the those items are subjected to one level of the row treatment and others are subjected to a different level of the row treatment.

The figure on the next page is designed to illustrate this idea and to suggest the way in which the issue it raises can be examined.

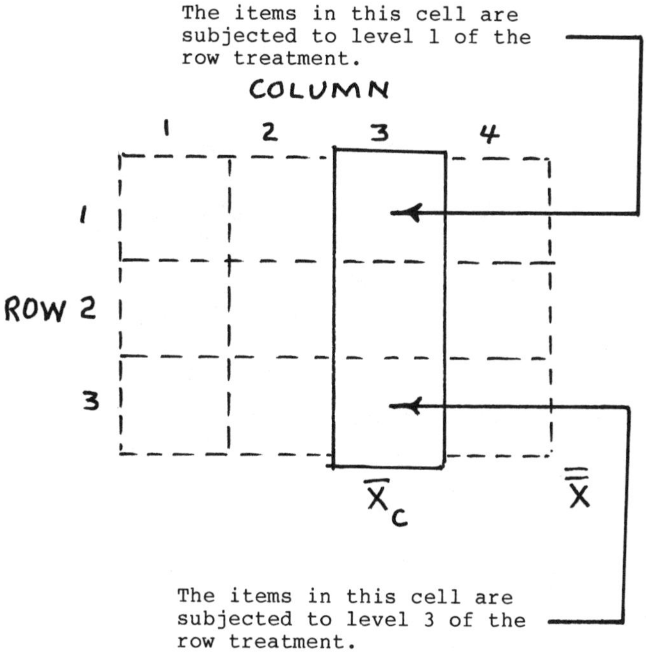
The items in this cell are subjected to level 1 of the row treatment.
COLUMN
1
2
3
4
1
ROW 2
3
$\bar{X}_c$
$\bar{\bar{X}}$
The items in this cell are subjected to level 3 of the row treatment.

If the row and column treatments are producing independent effects, the column treatment would add the same amount to every item in the column, regardless of the levels of the row treatments for those items.

One way to determine if this is in fact the case requires that we calculate the values of the observed cell means when they have been corrected by removing the observed row and column effects. When we make these corrections we do so under the assumption that the row and column effects are independent of each other. Thus we assume that the effect of a given column treatment is the same for every cell in the column and we also assume that the effect of a given row treatment is the same regardless of the level of the column treatment within that row.

If $\bar{X}_{rc}$ is an original (observed) cell mean in a given row, and if $(\bar{X}_r - \bar{\bar{X}})$ is the observed effect of that row treatment, the quantity $\bar{X}_{rc} - (\bar{X}_r - \bar{\bar{X}})$ will be the value of the cell mean corrected for the row effect.

Moreover, if we subtract $(\bar{X}_r - \bar{\bar{X}})$ from every cell mean in the rth row, the mean of the corrected means (in that row) will be $\bar{\bar{X}}$.

By a similar line of reasoning we can see that if $\bar{X}_{rc}$ is a given cell mean and if $(\bar{X}_c - \bar{\bar{X}})$ is the observed effect of the appropriate column treatment, the quantity $\bar{X}_{rc} - (\bar{X}_c - \bar{\bar{X}})$ will be the value of the cell mean corrected for that column effect. Moreover, if we correct the means of all of the other cells in the same column (by subtracting $(\bar{X}_c - \bar{\bar{X}})$ from each) the mean of the corrected means (in that column) will be $\bar{\bar{X}}$.

We will use the symbol $\bar{X}_I$ to represent a given cell mean that has been corrected for both the observed row and the observed column effect.

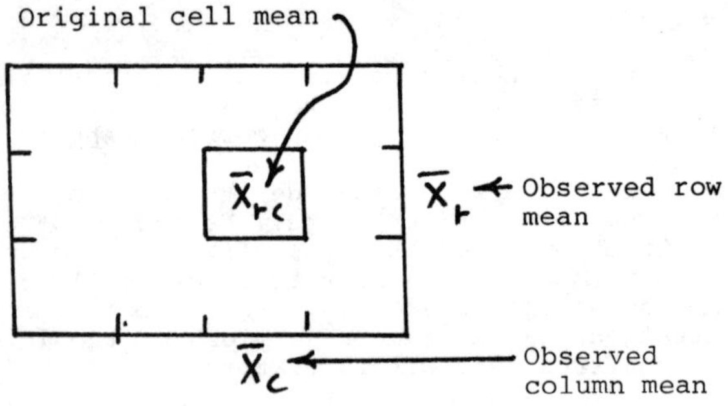

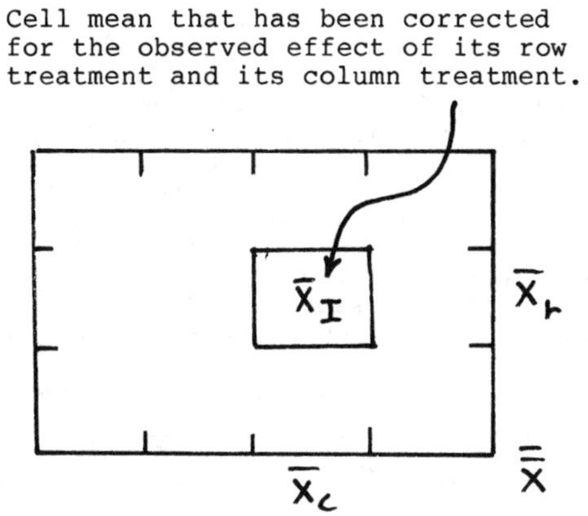

$$\bar{X}_I = \bar{X}_{rc} - (\bar{X}_r - \bar{\bar{X}}) - (\bar{X}_c - \bar{\bar{X}})$$

Observed row effect — Observed column effect

These calculations, applied to all RC of the cells, yield the third set of means that our analysis will employ.

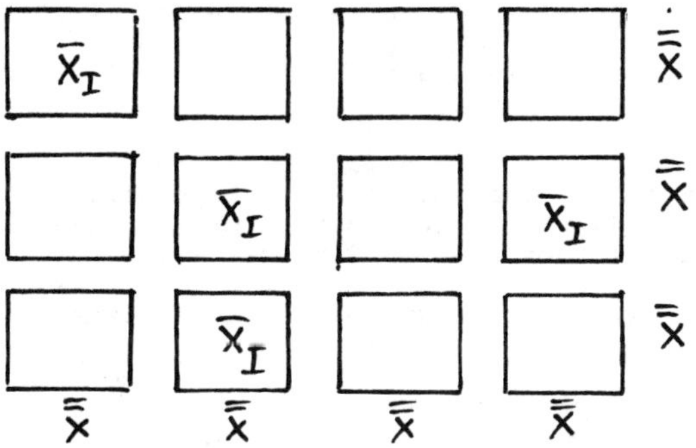

These means have the interesting property that the mean of any row of them will equal $\bar{\bar{X}}$ and this will also be true of the mean of any column. Moreover, the mean of all RC of them will also equal $\bar{\bar{X}}$.

If the student will reflect on how the several $\bar{X}_I$ s have been derived it will become clear that if our assumption of independent row and column effects is correct, the differences among the several $\bar{X}_I$ s must reflect nothing more than random sampling variations. But what if our assumption of independent effects is wrong? What if the effect of a given row treatment is interacting with the effect of the several column treatments to produce a different effect in each cell. Under these circumstances the quantity $(\bar{X}_r - \bar{\bar{X}})$ would be a reflection of an overall row effect, the quantity $(\bar{X}_c - \bar{\bar{X}})$ would reflect an effect an overall column effect and the quantity $(\bar{X}_I - \bar{\bar{X}})$ would reflect an effect on a given cell mean that could not be attributed to either the row or the column treatment per se. An effect of this sort is called an interaction.

Our problem in conducting our analysis of the data from a factorial experiment is to find a basis for deciding whether or not

1) our observed row effects are statistically reliable
2) our observed column effects are statistically reliable
3) our observed interaction effects are statistically reliable.

Consider now, how the data in a factorial study of this sort might appear if all of the observed differences among the various means merely reflected those variations that inevitably occur when one draws random samples from a single population. That is, consider the situation in which the row, column and the interaction effects are all zero.

Under those circumstances the quantity

$$\frac{nR\sum^{c}(\bar{X}_c - \bar{\bar{X}})^2}{C-1}$$ would be an unbiased estimate

of the variance of the original population just as the

quantity $\frac{n\sum^{K}(\bar{X} - \bar{\bar{X}})^2}{K - 1}$ was an estimate of σ_X^2 in

the one-way analysis discussed in Chapter 13.

In the present case, however, there would be C means instead of the K means in the one-way case, and rather than being based on n items (as in the one-way case) each column mean would be based on nR items.

In a similar fashion, the quantity

$$\frac{nC\sum^{R}(\bar{X}_r - \bar{\bar{X}})^2}{R - 1}$$

would also be an unbiased estimate of σ_X^2. Moreover, if all of the observed effects merely reflected random variations in sampling the

quantity $\frac{n\sum^{R}\sum^{C}(\bar{X}_I - \bar{\bar{X}})^2}{(R-1)(C-1)}$ would be a third

unbiased estimate of σ_X^2. The denominator of this statistic (R - 1)(C - 1) is its df. This quantity represents the number of cells in an R x C array that are free to vary under the restriction that every row and column mean must equal the grand mean.

There are, then, three independent estimates of σ_X^2 that can be derived from a factorial experiment when all of the samples come from a single population--as would be the case if the treatments and their interaction were all producing zero effects.

Of course, if the treatments and/or their interaction are not producing zero effects, that is they are actually producing differential effects, each of these estimates would be inflated by an amount that was proportional to those differential effects.

More specifically,

$$\frac{nR\sum^{C}(\bar{X}_c - \bar{\bar{X}})^2}{C - 1}$$

would be estimating σ_X^2 + (a quantity proportional to the actual effects of the column treatments)

$$\frac{nC\sum^{R}(\bar{X}_r - \bar{\bar{X}})^2}{R-1}$$ would be estimating σ_X^2 + (a quantity proportional to the actual effects of the row treatments)

and finally, the quantity

$$\frac{n\sum^{C}\sum^{R}(\bar{X}_I - \bar{\bar{X}})^2}{(R-1)(C-1)}$$ would be estimating σ_X^2 + (a quantity proportional to the actual effects of the interaction between the row and column treatments)

To decide whether a given one of these statistics is estimating σ_X^2 or whether it is estimating σ_X^2 plus a quantity proportional to the appropriate treatment effects, we compare it to a statistic that we can assume to be free of both the treatment effects and their interaction. If each of these effects is adding a quantity to every item on which they impinge they would not influence the variance within any of the cells. In other words, if (as we assume in the analysis of variance), the treatments are producing additive effects, the quantity

$$\frac{\sum^{R}\sum^{C}\sum^{n}(X - \bar{X}_{rc})^2}{CR(n-1)}$$

(the average of the s_X^2 within cells) would be an unbiased estimate of σ_X^2 that would be free of any of the treatment effects.

Accordingly, the statistic

$$F_{df=(C-1),\,(RC[n-1])} = \frac{\dfrac{nR\sum^{C}(\bar{X}_c - \bar{\bar{X}})^2}{C-1}}{\dfrac{\sum^{R}\sum^{C}\sum^{n}(X - \bar{X}_{rc})^2}{RC(n-1)}}$$

could be used to test the statistical hypothesis that the column treatments were producing zero effects.

Similarly, the statistic

$$F_{df=(R-1),(RC[n-1])} = \frac{\dfrac{nC\sum(\bar{X}_r - \bar{\bar{X}})^2}{R-1}}{\dfrac{\sum^{R}\sum^{C}\sum^{n}(X - \bar{X}_{rc})^2}{RC(n-1)}}$$

could be used to test for the statistical significance of the observed row effects.

And finally, the statistic

$$F_{df=(R-1)(C-1),(RC[n-1])} = \frac{\dfrac{n\sum^{R}\sum^{C}(\bar{X}_I - \bar{\bar{X}})^2}{(R-1)(C-1)}}{\dfrac{\sum^{R}\sum^{C}\sum^{n}(X - \bar{X}_{rc})^2}{RC(n-1)}}$$

could be used to test for the statistical significance of the observed interaction effects.

Here then, is the analysis of variance table that obtains from a two-way complete factorial design with n replications.

	SUM OF SQUARES	DF	MS	F
ROW	$nC\sum^{R}(\bar{X}_r - \bar{\bar{X}})^2$	$R-1$	$\frac{SS\ R}{DF\ R}$	$\frac{MS\ R}{MS\ W}$
COL	$nR\sum^{C}(\bar{X}_c - \bar{\bar{X}})^2$	$C-1$	$\frac{SS\ C}{DF\ C}$	$\frac{MS\ C}{MS\ W}$
INTER	$n\sum^{R}\sum^{C}(\bar{X}_I - \bar{\bar{X}})^2$	$(R-1)(C-1)$	$\frac{SS\ I}{DF\ I}$	$\frac{MS\ I}{MS\ W}$
SUB-TOT	$n\sum^{R}\sum^{C}(\bar{X}_{rc} - \bar{\bar{X}})^2$	$RC-1$		
WITHIN	$\sum^{R}\sum^{C}\sum^{n}(X - \bar{X}_{rc})^2$	$RC(n-1)$	$\frac{SS\ W}{DF\ W}$	
TOTAL	$\sum^{R}\sum^{C}\sum^{n}(X - \bar{\bar{X}})^2$	$RCn-1$		

Note: $\bar{X}_I = \bar{X}_{rc} - (\bar{X}_r - \bar{\bar{X}}) - (\bar{X}_c - \bar{\bar{X}})$

$$= \bar{X}_{rc} - \bar{X}_r - \bar{X}_c + 2\bar{\bar{X}}$$

thus $(\bar{X}_I - \bar{\bar{X}}) = \bar{X}_{rc} - \bar{X}_r - \bar{X}_c + \bar{\bar{X}}$

Here are the computing formulas for the same analysis.

	CONCEPTUAL	COMPUTATIONAL
ROW	$nC\sum^{R}(\bar{X}_r - \bar{\bar{X}})^2$	$\frac{1}{nC}\sum^{R} T_r^2 - \frac{T^2}{RCn}$
COL	$nR\sum^{C}(\bar{X}_c - \bar{\bar{X}})^2$	$\frac{1}{nR}\sum^{C} T_c^2 - \frac{T^2}{RCn}$
INTERACTION	$n\sum^{R}\sum^{C}(\bar{X}_I - \bar{\bar{X}})^2$	$\frac{1}{n}\sum^{R}\sum^{C} T_{rc}^2 - \frac{1}{nC}\sum T_r^2 - \frac{1}{nR}\sum^{C} T_c^2 - \frac{T^2}{RCn}$
SUB-T	$n\sum^{R}\sum^{C}(\bar{X}_{rc} - \bar{\bar{X}})^2$	$\frac{1}{n}\sum^{R}\sum^{C} T_{rc}^2 - \frac{T^2}{RCn}$
WITHIN	$\sum^{R}\sum^{C}\sum^{n}(X - \bar{X}_{rc})^2$	$\sum^{R}\sum^{C}\sum^{n} X^2 - \frac{1}{n}\sum^{R}\sum^{C} T_{rc}^2$
TOTAL	$\sum^{R}\sum^{C}\sum^{n}(X - \bar{\bar{X}})^2$	$\sum^{R}\sum^{C}\sum^{n} X^2 - \frac{T^2}{RCn}$

$T_{rc} = n\bar{X}_{rc}$ = the cell total

$T_r = nC\bar{X}_r$ = the row total

$T_c = nR\bar{X}_c$ = the colomn total

$T = nRC\bar{\bar{X}}$ = the grand total

A careful examination of the tables for the two-way analysis of variance can reveal certain features of this powerful statistical tool. For example, notice that the degrees of freedom for rows, columns and interaction when added together equals the degrees of freedom for the subtotal. Furthermore, when one adds the degrees of freedom for the subtotal to the degrees of freedom for the within sum of squares, one obtains the degrees of freedom for the total sum of squares.

It is of interest that a similar set of relationships prevail for the sums of squares themselves. If one adds together the sums of squares for rows, columns and interaction, one obtains the sum of squares for the subtotal. If one then adds the within sum of squares to the subtotal sum of squares, one obtains the total sum of squares. These relationships are not at all obvious when one examines the several conceptual formulae, but they stand out clearly when one examines the computational formulae (on page 269).

A numerical example of the two-way analysis of variance

We can gain some appreciation of the considerable versatility of the two-way analysis of variance if we examine its numerical application to the study that was described at the beginning of this chapter.

In that study each of 60 subjects was tested for accuracy at throwing darts at a target under an experimental arrangement in which the test room was set at 1 of 3 possible illumination levels and 4 possible temperatures. With this design there were 12 possible combinations of test conditions and there were 5 subjects randomly assigned to each combination of conditions.

The next page shows the kind of data that might have been produced in a study of this sort. Each number in the body of the array is a single subject's score (out of a possible perfect 25). Also shown are the $\bar{X}$s for each cell, as well as the means of the rows and of the columns.

TEMPERATURE

ILLUMINATION		40°	50°	60°	70°
	BRIGHT	8 7 7 15 14	11 13 12 15 15	13 19 20 19 18	19 15 21 14 21
	NORMAL	10 8 7 8 7	11 10 12 12 12	9 11 16 16 11	11 17 13 12 12
	DIM	5 5 6 2 7	7 6 10 6 10	12 12 9 7 10	15 14 12 12 7

Cell Mean — Row Mean

	40°	50°	60°	70°	Row Mean
BRI	10.2	13.2	17.8	18.0	14.80
NOR	8.00	11.4	12.6	13.0	11.25
DIM	5.00	7.8	12.6	12.0	8.70
Col Mean	7.7333	10.80	13.4567	14.3333	11.58 (Grand mean)

These means deserve careful consideration because they provide the answers to the questions raised in the study. Briefly stated, that study asked how accuracy at dart throwing might be influenced by room illumination as well as by room temperature. The next three figures provide a graphic representation of the major trends in its results.

Figure A shows the mean, across rows, for each column; thus Figure A provides a graphic description of the overall effect of temperature on accuracy. Figure B shows the means, across columns, for each row; thus Figure B provides a graphic description of the overall effect of illumination level on accuracy. Figure C shows the means obtained under each combination of room illumination and room temperature; thus Figure C provides a graphic account of the manner in which the two treatment variables have jointly influenced accuracy.

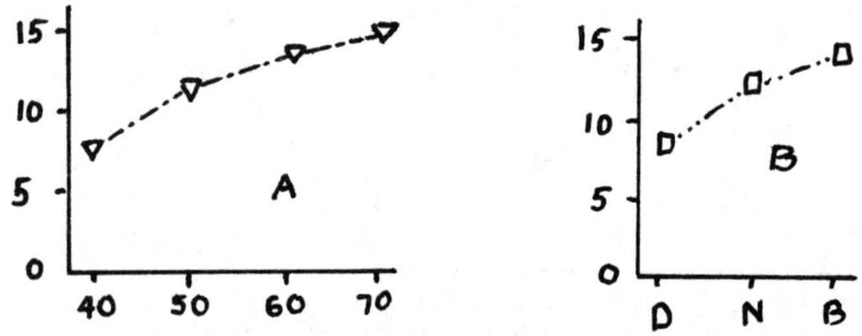

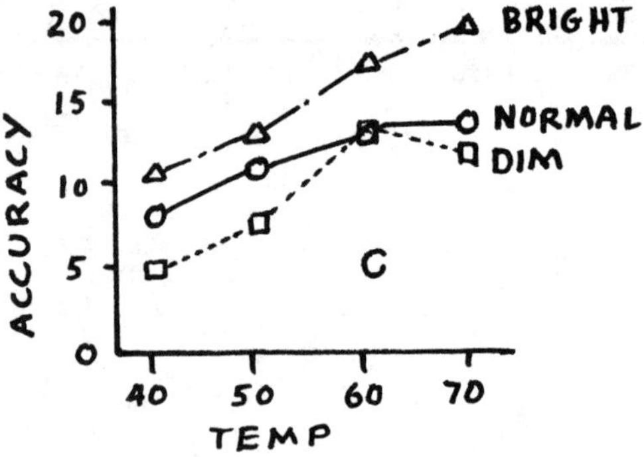

Examination of the figures on the previous page reveals that increases in both room illumination and room temperature were accompanied by an increase in accuracy. Moreover, since the three functions in Figure C appear to be approximately parallel, the data also indicate that the two factors (illumination and temperature) had largely independent effects. Another way to say this is to assert that the effects of temperature (an increase in accuracy with an increase in temperature) were the same regardless of the level of room illumination. A third way to say the same thing is to assert that the data provide no evidence of an interaction between the effects of room temperature and room illumination.

Of course in this study (as in any other study one might do) summarizing the data and seeking to understand their implications is only a first step in the analysis of its results. The fact remains that as long as one's data represent measures on samples (rather than on populations) and as long as these measures exhibit some random variability, one must also cope with the possibility that any trends that they might also exhibit (including those seen in Figures A, B, and C) are merely an expression of that random variability.

The next page shows the results of the analysis of variance that we conduct to resolve this possibility. As seen in the Analysis of Variance Table, the Fs for both the row means and the column means (29.10 and 20.58 respectively) are both sufficiently large that they fall among those extreme values of F that would only occur by chance .01 of the time when treatments have no effect. Accordingly we reject the null hypothesis at the α = .01 level of significance and conclude that the observed increases in accuracy as temperature and illumination increased did not occur by chance (i.e., we conclude that these effects are a product of the two treatment variables.) Moreover, since we failed to obtain a statistically significant F in the test for interaction, we also conclude that the two treatment variables do not interact with each other.

	Sum of Squares	df	Mean Square	F
Rows	375.433	2	187.717	29.1033
Columns	398.184	3	132.728	20.5779
Interaction	29.3667	6	4.89444	.758826
Subtotal	802.984	11		
Within	309.601	48	6.45002	
Total	1112.58	59		

We can see why these conclusions are justified if we examine the various mean squares that contribute to each of the three Fs in the analysis of variance. Consider first the Mean Square for Within. This quantity serves as the denominator of all three Fs. Earlier we learned that the Mean Square Within is the average of the S_x^2 s that are calculated on the scores of subjects that are exposed to a given combination of the two treatment variables. Shown below are each of the 12 S_x^2 s calculated on the 5 scores that are in each cell.

15.7001	3.20007	7.70026	11
1.50002	.799957	10.2999	5.49997
3.5	4.19995	4.50001	9.50003

The average of these 12 S_x^2 s is 6.45002 which is the same value one obtains when one divides the Sum of Squares for Within (i.e., 304.601) by the Degrees of Freedom for Within (i.e., 48).

As noted earlier, when the null hypothesis is true the average of the S_x^2 s is an unbiased estimate of the variance of the population from which the samples were drawn. Moreover, because it is a measure of variability around the obtained means of each cell, it is an estimate that will be free of any treatment effects. For this reason, the average of the S_x^2 s (i.e., the MS Within) is also an unbiased estimate of σ_x^2 when the null hypothesis is false and the treatments have, in fact, non zero effects.

The next quantity to examine is the numerator of the F for Rows. This quantity (i.e., 187.717) consists of the value one obtains when one divides the Sum of Squares for Rows (i.e., 375.433) by the Degrees of Freedom for Rows (i.e., 2). Another way to calculate the Mean Square for Rows is to calculate the $s_{\bar{X}}^2$ for rows and then to multiply this quantity by the number of items that contribute to each of the the row means. In the present case $s_{\bar{X}}^2$ for rows is 9.38583 and each row mean is based on nC = 20 items). If we multiply 9.38583 by 20 we obtain 187.717 which is the Mean Square for Rows. As discussed earlier, in an analysis of variance, the several Mean Squares (in the present case the Mean Squares for Rows and for columns and for interaction as well) are unbiased estimates of σ_x^2 when the null hypothesis is true, and they are all unbiased estimates of σ_x^2 plus a quantity that is proportional to the differential magnitude of the appropriate effects, when the null hypothesis is false. In the example we are studying the MS for Rows is almost 30 times as large as the MS Within. (Notice that the F for Rows equals 29.1033.) It is this observation that convinces us to reject the null hypothesis with respect to rows and to conclude that the row treatments have yielded non zero effects.

Similar algebraic and arithmetic relationships exist for the Mean Square for Columns and the conclusions we draw from them entail a similar logic. The MS for columns is equal to the Sum of Squares for Columns divided by the Degrees of Freedom for Columns.

re: $\text{MS Columns} = 132.728 = \frac{398.184}{3}$

Moreover, like the Mean Square for Rows, the Mean Square for Columns can be calculated in a direct fashion. For these data the $s_{\bar{X}}^2$ for columns is 8.84852 and each column mean contains nR = 5 x 3 = 15 data points. If we multiply 8.84852 by 15 we obtain 132.728 which is the Mean Square for Columns.

Calculations of the Mean Square for Interaction can also proceed in either of two ways. If we have already calculated the Sums of Square for Rows and Columns and have either calculated the Sum of Squares for subtotal or the Sum of Square for Within as well as the total Sum of Squares, we can derive the Sum of Squares for Interaction by a subtraction procedure.

That is, we take advantage of the relationships revealed in the computing formulae, such that

SS Interaction = SS Subtotal - SS Rows - SS Column

or

SS Interaction = SS Total - SS Within - SS Rows - SS Column

A more tedious, but more intuitively reasonable, way to calculate the MS for Interaction is to do so directly. The first step is to correct each of the cell means for the observed row and column effects. For example, the mean for the cell in Row 2, Column 3 is 12.6. The row effect for the cells in Row 2 is numerically equal to the mean for Row 2 minus the grand mean,

i.e., 11.25 - 11.5833 = 1.3333 = Effect for Row 2

The column effect for the cells in Column 3 is numerically equal to the mean for Column 3 minus the grand mean,

i.e., 13.4667 - 11.5833 = 1.8834 = Effect for Col 3

If we now subtract the row 2 and the column 3 effect from the mean of the cell in Row 2, Column 3, we will obtain a new mean (i.e., an $\bar{X}_I$) that has been corrected for the observed row and column effects

i.e., 12.6000 - (-.3333) - 1.8834 = 11.05 = $\bar{X}_I$

If we proceed to calculate the values of the 11 other corrected cell means (by subtracting the appropriate row and column effects in each case), we will obtain the following set of corrected cell means (i.e., values of $\bar{X}_I$).

10.8333	10.7667	12.7	12.0333
12.1833	12.5167	11.05	10.5833
11.7333	11.4667	11	12.1333

As we can easily check for ourselves, this array of corrected cell means displays the characteristic that every row and every column has a mean that (within rounding error) equals the grand mean (i.e., 11.5833). If we calculate the $s_{\bar{x}}^2$ for the corrected cell means and if its value is multiplied by 5 (the number of items on which each of the corrected cell means is based) we will obtain the value of the Mean Square for Interaction. That is, we will obtain this value provided that we use the correct degrees of freedom in performing these calculations.

In the present numerical example the correct degrees of freedom is 6 (not 11 as one might initially think). To see why we must recognize that as seen above when we correct the cell means for observed row and and column effects, we automatically arrange that the mean of every row as well as of every column will be equal to the grand mean.

In doing so, we use some of the information in the cell means and this causes us to lose degrees of freedom. In general, the degrees of freedom in a set of corrected cell means is the number of these means that are free to assume any value whatsoever. This is not a new idea. It will be recalled that the statistic

$$S_X^2 = \frac{\sum(X - \bar{X})^2}{n - 1}$$

has n - 1 rather than n degrees of freedom because in any set of n values of $(X - \bar{X})^2$ the final $(X - \bar{X})$ must have a value such that $\sum(X - \bar{X}) = 0$

In the present set of circumstances we have RC = 12 corrected cell means, but in fact only (R - 1)(C - 1) = 2 x 3 = 6 of the values of $(\bar{X}_I - \bar{\bar{X}})^2$ can assume any value whatsoever. To see why, consider that while each row contains C corrected cell means, only C - 1 of these values are free to vary; the last value in a row must always be the one value that will cause the mean of the items in the row to exactly equal the grand mean. Similarly each column will contain only R - 1 items that are free to vary. The last item in each column must be such that the mean of the column will be exactly equal to the grand mean. These two facts considered together imply that in this 3 x 4 array there are only 2 x 3 = 6 df for the corrected cell means.

If we calculate the sum of the squared deviations of the several corrected cell means from the grand mean we obtain

$$\sum^{R}\sum^{C}(\bar{X}_I - \bar{\bar{X}})^2 = 5.87334$$

If we then divide this quantity by df = 6 we obtain .97889 which is the value of $s^2_{\bar{X}_I}$. If we then multiply this quantity by 5 (the number of items on which each corrected cell mean is based, we arrive at the value of the Mean Square for Interaction.

i.e., MS Interaction = 5(.97889) = 4.89445

Since this value is the same (within rounding errors) as the value obtained using the indirect subtraction procedure described earlier, we can be confident that our calculations are correct.

A Final Comment

By indicating that the observed differences among the row and the column means are larger than expected by chance, we are led to conclude that those differences are due to the treatments. Without the objective evidence provided by the analysis of variance, this conclusion might have little if any justification.

In addition to providing a basis for the conclusions we draw, statistical considerations can aid in the initial phases of an investigation by pointing to the kind of factors that will require attention. For example, in the dart throwing experiment we have been studying, the concept of interaction makes it clear why it was important to test all possible combinations of treatments.

Statistical considerations also help in finding ways to summarize and arrange data. Without the framework provided by the statistical analysis, the aspects of data that received attention in the dart throwing experiment might have been determined by our preconceptions and biases instead of by the formal aspects of the work.

This is not to suggest that statistics can take the place of good intuitions and creative ideas when it comes to doing effective research. By the same token, however, we should recognize that good intuitions and creative ideas are no substitute for the objective quantitative approach provided by an appropriate statistical analysis. These considerations make it easy to see why a thorough knowledge of statistics is one of the most important of the intellectual tools available to the contemporary investigator. It is hoped that by illuminating the logic of statistical reasoning, this book will facilitate the intelligent use of these tools.

APPENDIX I: Statistical Tables

Instructions on how to read these tables are provided on the next few pages. As noted at several points in this text, the sampling distributions on which these tables are based are all contained in (i.e., are special cases of) the F distribution. Notice, for example, that Table V tells us that when df = ∞ , a value of t = 1.960 cuts off the upper .025 of the sampling distribution for t. This agrees with the observation from Table I that (.5 - .025) = .4750 of a normal distribution lies between the mean and an ordinate at Z = 1.96. When df = 6, on the other hand, t must equal 2.447 if it is to cut off the upper .025 of the sampling distribution for t (see Table V). Notice also that when t has 6 df, $t^2 = 2.447^2 =$ 5.99. As seen in Table III, 5.99 is the value of F that cuts off the upper .05 of the sampling distribution for F, when df = 1 and 6.

In a similar vain, Table II tells us that when df = 6, a value of $\frac{s_x^2}{\sigma_x^2}$ = 2.10 cuts off the upper .05 of the sampling distribution for $\frac{s_x^2}{\sigma_x^2}$. As seen in Table III, when df = 6, ∞ the value of F that cuts off the upper .05 of the F distribution is 2.10. Finally, as seen in Table VI, when df = 6, the value of χ^2 that cuts off the upper .05 of the χ^2 distribution is 12.592. Notice that 12.592/6 is (within rounding error) equal to 2.10.

These observations can be summarized by asserting that if we let the symbol ν represent a given degrees of freedom, then, at a given level of α :

1) $Z_x \quad = \quad t_{df=\infty}$

2) $F_{df=1,\nu} \quad = \quad t^2_{df=\nu}$ (for a two-tailed test)

3) $\frac{s_x^2}{\sigma_x^2}\,{}_{df=\nu} \quad = \quad F_{df=\nu,\infty}$

4) $\frac{\chi^2_{df=\nu}}{\nu} \quad = \quad \frac{s_x^2}{\sigma_x^2}\,{}_{df=\nu}$

TABLE I - Areas under the Normal Curve

Entries in the body of Table I show the proportion of the total area in a normal distribution with a mean of 0 and a $\sigma = 1$ that lies between the mean and an ordinate erected at the value of Z indicated at the margins of the table. For example, .3925 of the total area in such a distribution lies between the mean and a value of Z = 1.24.

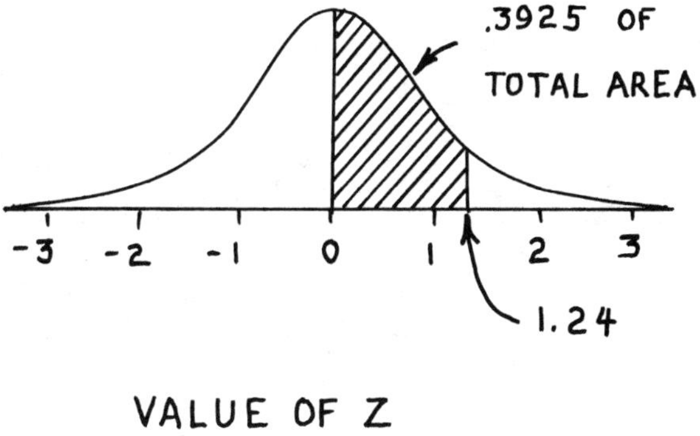

TABLE I Areas Under the Normal Curve

	.00	.01	.02	.03	.04	.05	.06	.07	.08	.09
0.0	.0000	.0040	.0080	.0120	.0160	.0199	.0239	.0279	.0319	.0359
0.1	.0398	.0438	.0478	.0517	.0557	.0596	.0636	.0675	.0714	.0753
0.2	.0793	.0832	.0871	.0910	.0948	.0987	.1026	.1064	.1103	.1141
0.3	.1179	.1217	.1255	.1293	.1331	.1368	.1406	.1443	.1480	.1517
0.4	.1554	.1591	.1628	.1664	.1700	.1736	.1772	.1808	.1844	.1879
0.5	.1915	.1950	.1985	.2019	.2054	.2088	.2123	.2157	.2190	.2224
0.6	.2257	.2291	.2324	.2357	.2389	.2422	.2454	.2486	.2517	.2549
0.7	.2580	.2611	.2642	.2673	.2704	.2734	.2764	.2794	.2823	.2852
0.8	.2881	.2910	.2939	.2967	.2995	.3023	.3051	.3078	.3106	.3133
0.9	.3159	.3186	.3212	.3238	.3264	.3289	.3315	.3340	.3365	.3389
1.0	.3413	.3438	.3461	.3485	.3508	.3531	.3554	.3577	.3599	.3621
1.1	.3643	.3665	.3686	.3708	.3729	.3749	.3770	.3790	.3810	.3830
1.2	.3849	.3869	.3888	.3907	.3925	.3944	.3962	.3980	.3997	.4015
1.3	.4032	.4049	.4066	.4082	.4099	.4115	.4131	.4147	.4162	.4177
1.4	.4192	.4207	.4222	.4236	.4251	.4265	.4279	.4292	.4306	.4319
1.5	.4332	.4345	.4357	.4370	.4382	.4394	.4406	.4418	.4429	.4441
1.6	.4452	.4463	.4474	.4484	.4495	.4505	.4515	.4525	.4535	.4545
1.7	.4554	.4564	.4573	.4582	.4591	.4599	.4608	.4616	.4625	.4633
1.8	.4641	.4649	.4656	.4664	.4671	.4678	.4686	.4693	.4699	.4706
1.9	.4713	.4719	.4726	.4732	.4738	.4744	.4750	.4756	.4761	.4767
2.0	.4772	.4778	.4783	.4788	.4793	.4798	.4803	.4808	.4812	.4817
2.1	.4821	.4826	.4830	.4834	.4838	.4842	.4846	.4850	.4854	.4857
2.2	.4861	.4864	.4868	.4871	.4875	.4878	.4881	.4884	.4887	.4890
2.3	.4893	.4896	.4898	.4901	.4904	.4906	.4909	.4911	.3913	.1916

Normal Curve (cont'd)

	.00	.01	.02	.03	.04	.05	.06	.07	.08	.09
2.4	.4918	.4920	.4922	.4925	.4927	.4929	.4931	.4932	.4934	.4936
2.5	.4938	.4940	.4941	.4943	.4945	.4946	.4948	.4949	.4951	.4952
2.6	.4953	.4955	.4956	.4957	.4959	.4960	.4961	.4962	.4963	.4964
2.7	.4965	.4966	.4967	.4968	.4969	.4970	.4971	.4972	.4973	.4974
2.8	.4974	.4975	.4976	.4977	.4977	.4978	.4979	.4979	.4980	.4981
2.9	.4981	.4982	.4982	.4983	.4984	.4984	.4985	.4985	.4986	.4986
3.0	.4987	.4987	.4987	.4988	.4988	.4989	.4989	.4989	.4990	.4990
3.1	.49903									
3.2	.49931									
3.3	.49952									
3.4	.49966									
3.5	.49977									

TABLE II - Critical Values of $\frac{s_x^2}{\sigma_x^2}$

Entries in the body of Table II show (for a given df) the value of $\frac{s_x^2}{\sigma_x^2}$ that cuts off the proportion of the sampling distribution of this statistic that is indicated at the head of the column. For example, when df = 3, a value of $\frac{s_x^2}{\sigma_x^2}$ = 2.60 cuts off the upper .05 of the sampling distribution of this statistic.

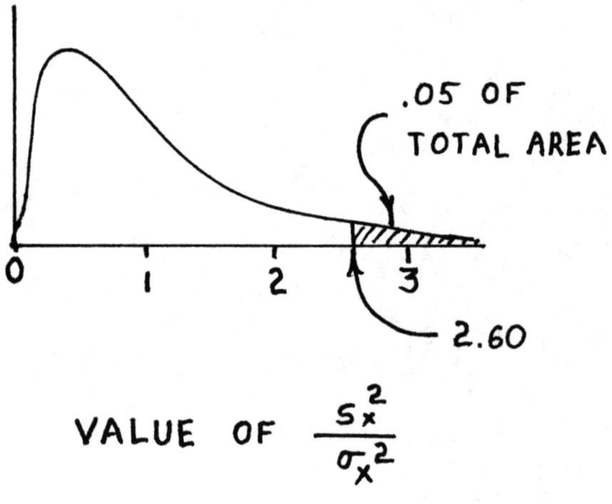

TABLE II

Critical Values of the $\frac{s_x^2}{\sigma_x^2}$ Distribution

df	Lower .025	Upper .025	Lower .005	Upper .005	Upper .05	Upper .01
1	.001	5.02	.00004	7.88	3.84	6.63
2	.025	3.69	.005	5.30	3.00	4.61
3	.072	3.12	.024	4.28	2.60	3.78
4	.121	2.79	.052	3.72	2.37	3.32
5	.166	2.57	.082	3.35	2.21	3.02
6	.206	2.41	.113	3.09	2.10	2.80
7	.241	2.29	.141	2.90	2.01	2.64
8	.272	2.19	.168	2.74	1.94	2.51
9	.300	2.11	.193	2.62	1.88	2.41
10	.325	2.05	.216	2.52	1.83	2.32
20	.480	1.71	.372	2.00	1.57	1.88
30	.560	1.57	.460	1.79	1.46	1.70
40	.611	1.48	.518	1.67	1.39	1.59
50	.645	1.43	.559	1.59	1.35	1.52
60	.675	1.39	.592	1.53	1.32	1.47
100	.741	1.30	.671	1.40	1.24	1.36
120	.763	1.27	.699	1.36	1.22	1.32
200	.813	1.21	.762	1.28	1.17	1.25
500	.878	1.13	.843	1.17	1.11	1.15
∞	1.000	1.00	1.000	1.00	1.00	1.00

This figure is adapted with the permission of McGraw Hill Book Company from Introduction to Statistical Analysis by Wilfred J. Dixon and Frank J. Massey, Jr. Copyright 1969.

TABLE III = Critical Values of the F Distribution
(α = .05)

Entries in the body of the table are the values that cut off the upper .05 of the sampling distribution of $F = \frac{s_1^2}{s_2^2}$. The headings for the columns and the rows are the degrees of freedom for s_1^2 and s_2^2 respectively. For example, when df = 4 for s_1^2 and 21 for s_2^2, a value of $F = \frac{s_1^2}{s_2^2}$ = 2.71 cuts off the upper .05 of the sampling distribution.

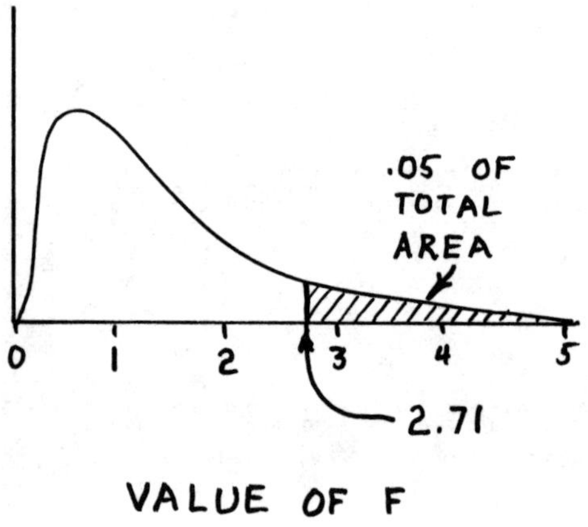

TABLE III Critical Values of the F Distribution (α = .05)

	1	2	3	4	5	6	7	8	9	10	15	20	∞
1	161	200	216	225	230	234	237	239	241	242	246	248	254
2	18.5	19.0	19.2	19.2	19.3	19.3	19.4	19.4	19.4	19.4	19.4	19.4	19.5
3	10.1	9.55	9.28	9.12	9.01	8.91	8.89	8.85	8.81	8.79	8.70	8.66	8.53
4	7.71	6.94	6.59	6.39	6.26	6.16	6.09	6.04	6.00	5.96	5.86	5.80	5.63
5	6.61	5.79	5.41	5.19	5.05	4.95	4.88	4.82	4.77	4.74	4.62	4.56	4.37
6	5.99	5.14	4.76	4.53	4.39	4.28	4.21	4.15	4.10	4.06	3.94	3.87	3.76
7	5.59	4.74	4.35	4.12	3.97	3.87	3.79	3.73	3.68	3.64	3.51	3.44	3.23
8	5.32	4.46	4.07	3.84	3.69	3.58	3.50	3.44	3.39	3.35	3.22	3.15	2.93
9	5.12	4.26	3.86	3.63	3.48	3.37	3.29	3.23	3.18	3.14	3.01	2.94	2.71
10	4.96	4.10	3.71	3.48	3.33	3.22	3.14	3.07	3.02	2.98	2.85	2.77	2.54
11	4.84	3.98	3.59	3.36	3.20	3.09	3.01	2.95	2.90	2.85	2.72	2.65	2.40
12	4.75	3.89	3.49	3.26	3.11	3.00	2.91	2.85	2.80	2.75	2.62	2.54	2.30
13	4.67	3.81	3.41	3.18	3.03	2.92	2.83	2.77	2.71	2.67	2.53	2.46	2.21
14	4.60	3.74	3.31	3.11	2.96	2.85	2.76	2.70	2.65	2.60	2.46	2.39	2.13
15	4.54	3.68	3.29	3.06	2.90	2.79	2.71	2.64	2.59	2.54	2.40	2.33	2.07
16	4.49	3.63	3.24	3.01	2.85	2.74	2.66	2.59	2.54	2.49	2.35	2.28	2.01
17	4.45	3.59	3.20	2.96	2.81	2.70	2.61	2.55	2.49	2.45	2.31	2.23	1.96
18	4.41	3.55	3.16	2.93	2.77	2.66	2.58	2.51	2.46	2.41	2.27	2.19	1.92
19	4.38	3.52	3.13	2.90	2.74	2.63	2.54	2.48	2.42	2.38	2.23	2.16	1.88
20	4.35	3.49	3.10	2.87	2.71	2.60	2.51	2.45	2.39	2.35	2.20	2.12	1.84
21	4.32	3.47	3.07	2.84	2.68	2.57	2.49	2.42	2.37	2.32	2.18	2.10	1.81
22	4.30	3.44	3.05	2.82	2.66	2.55	2.46	2.40	2.34	2.30	2.15	2.07	1.78
23	4.28	3.42	3.03	2.80	2.64	2.53	2.44	2.37	2.32	2.27	2.13	2.05	1.76
24	2.26	3.40	3.01	2.78	2.62	2.51	2.42	2.36	2.30	2.25	2.11	2.03	1.73
25	4.24	3.39	2.99	2.76	2.60	2.49	2.40	2.34	2.28	2.24	2.09	2.01	1.71
30	4.17	3.32	2.92	2.69	2.53	2.42	2.33	2.27	2.21	2.16	2.01	1.93	1.62
∞	3.84	2.00	2.60	2.37	2.21	2.10	2.01	1.94	1.88	1.83	1.67	1.57	1.00

TABLE IV - Critical Values of the F Distribution
$(\alpha = .01)$

Entries in the body of the table are the values that cut off the upper .01 of the sampling distribution of $F = \frac{s_1^2}{s_2^2}$. The headings for the columns and rows are the degrees of freedom for s_1^2 and s_2^2 respectively. For example, when df = 4 for s_1^2 and 21 for s_2^2, a value of $F = \frac{s_1^2}{s_2^2} = 4.37$ cuts off the upper .01 of the sampling distribution.

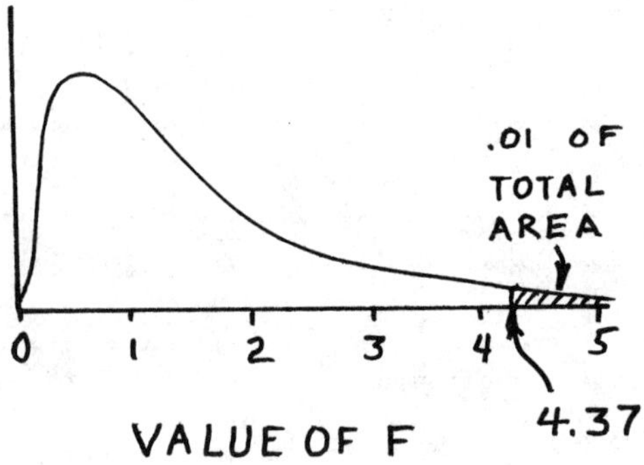

TABLE IV Critical Values of the F Distribution (α = .01)

	1	2	3	4	5	6	7	8	9	10	15	20	∞
1	4,042	5,000	5,403	5,625	5,764	5,859	5,928	5,928	6,023	6,056	6,157	6,209	6,366
2	98.5	99.0	99.2	99.2	99.3	99.3	99.4	99.4	99.4	99.4	99.4	99.4	99.5
3	34.1	30.8	29.5	28.7	28.2	27.9	27.7	27.5	27.3	27.2	26.9	26.7	26.1
4	21.2	18.0	16.7	16.0	15.5	15.2	15.0	14.8	14.7	14.5	14.2	14.0	13.5
5	16.3	13.3	12.1	11.4	11.0	10.7	10.5	10.3	10.2	10.1	9.72	9.55	9.02
6	14.7	10.9	9.78	9.15	8.75	8.47	8.26	8.10	7.98	7.87	7.56	7.40	6.88
7	12.2	9.55	8.45	7.85	7.46	7.19	6.99	6.84	6.72	6.62	6.31	6.16	5.65
8	11.3	8.65	7.59	7.01	6.63	6.37	6.18	6.03	5.91	5.81	5.52	5.36	4.86
9	10.6	8.02	6.99	6.42	6.06	5.80	5.61	5.47	5.35	5.26	4.96	4.81	4.31
10	10.0	7.56	6.55	5.99	5.64	5.39	5.20	5.06	4.91	4.85	4.56	4.41	3.91
11	9.65	7.21	6.22	5.67	5.32	5.07	4.89	4.74	4.63	4.54	4.25	4.10	3.60
12	9.33	6.93	5.95	5.41	5.06	4.82	4.64	4.50	4.39	4.30	4.01	3.86	3.36
13	9.07	6.70	5.74	5.21	4.86	4.62	4.44	4.30	4.19	4.10	3.82	3.66	3.17
14	8.86	6.51	5.56	5.04	4.70	4.46	4.28	4.14	4.03	3.94	3.66	3.51	3.00
15	8.68	6.36	5.42	4.89	4.56	4.32	4.14	4.00	3.89	3.80	3.52	3.37	2.87
16	8.53	6.23	5.29	4.77	4.44	4.20	4.03	3.89	3.78	3.69	3.41	3.26	2.75
17	8.40	6.11	5.19	4.67	4.34	4.10	3.93	3.70	3.68	3.59	3.31	3.16	2.65
18	8.29	6.01	5.09	4.58	4.25	4.01	3.84	3.71	3.60	3.51	3.23	3.08	2.57
19	8.19	5.93	5.01	4.50	4.17	3.94	3.77	3.63	3.52	3.43	3.15	3.00	2.49
20	8.10	5.85	4.94	4.43	4.10	3.87	3.70	3.56	3.46	3.37	3.09	2.94	2.42
21	8.02	5.78	4.87	4.37	4.04	3.81	3.64	3.51	3.40	3.31	3.03	2.88	2.36
22	7.95	5.72	4.82	4.31	3.99	3.76	3.59	3.45	3.35	3.26	2.98	2.83	2.31
23	7.88	5.66	4.76	4.26	3.94	3.71	3.54	3.41	3.30	3.21	2.93	2.78	2.26
24	7.82	5.61	4.72	4.22	3.90	3.67	3.50	3.36	3.26	3.17	2.89	2.74	2.21
25	7.77	5.57	4.68	4.18	3.86	3.63	3.46	3.32	3.22	3.13	2.85	2.70	2.17
30	7.56	5.39	4.51	4.02	3.70	3.47	3.30	3.17	3.07	2.98	2.70	2.55	2.01
∞	6.63	4.61	3.78	3.32	3.02	2.80	2.64	2.51	2.41	2.32	2.04	1.88	1.00

TABLE V - Critical Values of the t Distribution

Entries in the body of Table V are the values of t that cut off the proportion of the sampling distribution of t indicated by the headings at the top of the columns. The row headings indicate the degrees of freedom. For example, when t has 11 degrees of freedom a value of t = ± 2.201 cuts off the upper and lower .025 of the distribution. This then is the critical value for a two-tailed test with α = .05. Notice that for a one-tailed test at the same α level, the critical value for t with 11 degrees of freedom is 1.796.

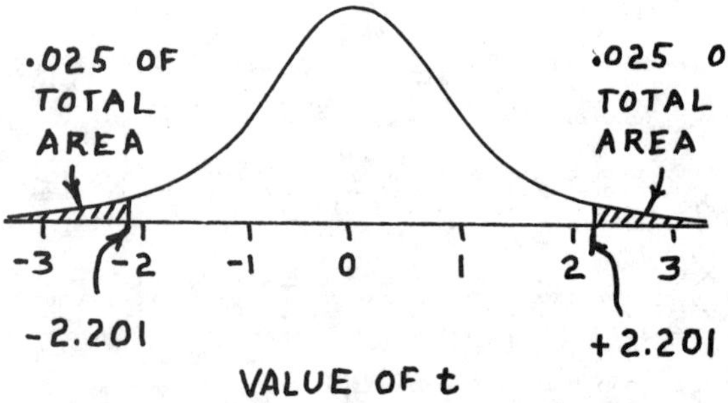

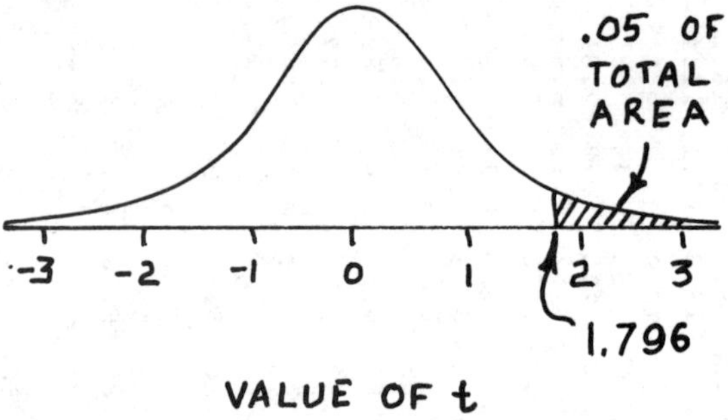

TABLE V Critical Values of the t Distribution

df	Level of Significance for One-Tailed Test (α)					
	0.10	0.05	0.025	0.01	0.005	0.0005
	Level of Significance for Two-Tailed Text (α)					
	0.20	0.10	0.05	0.02	0.01	0.001
1	3.078	6.314	12.706	31.821	63.657	636.619
2	1.886	2.920	4.303	6.965	9.925	31.598
3	1.638	2.353	3.182	4.541	5.841	12.941
4	1.533	2.132	2.776	3.747	4.604	8.610
5	1.476	2.015	2.571	3.365	4.032	6.859
6	1.440	1.943	2.447	3.143	3.707	5.959
7	1.415	1.895	2.365	2.998	3.499	5.405
8	1.397	1.860	2.306	2.896	3.355	5.041
9	1.383	1.833	2.262	2.821	3.250	4.781
10	1.372	1.812	2.228	2.764	3.169	4.587
11	1.363	1.796	2.201	2.718	3.106	4.437
12	1.356	1.782	2.179	2.681	3.055	4.318
13	1.350	1.771	2.160	2.650	3.012	4.221
14	1.345	1.761	2.145	2.624	2.977	4.140
15	1.341	1.753	2.131	2.602	2.947	4.073
16	1.337	1.746	2.120	2.583	2.921	4.015
17	1.333	1.740	2.110	2.567	2.898	3.965
18	1.330	1.734	2.101	2.552	2.878	3.992
19	1.328	1.729	2.093	2.539	2.861	3.883
20	1.325	1.725	2.086	2.528	2.845	3.850
21	1.323	1.721	2.080	2.518	2.831	3.819
22	1.321	1.717	2.074	2.508	2.819	3.792
23	1.319	1.714	2.069	2.500	2.807	3.767
24	1.318	1.711	2.064	2.492	2.797	3.745
25	1.316	1.708	2.060	2.485	2.787	3.725
26	1.315	1.706	2.056	2.479	2.779	3.707
27	1.314	1.703	2.052	2.473	2.771	3.690
28	1.313	1.701	2.048	2.467	3.763	3.674
29	1.311	1.699	2.045	2.462	2.756	3.659
30	1.310	1.697	2.042	2.457	2.750	3.646
40	1.303	1.684	2.021	2.423	2.704	3.551
60	1.296	1.671	2.000	2.390	2.660	3.460
120	1.289	1.658	1.980	2.358	3.617	3.617
∞	1.282	1.645	1.960	2.326	2.576	3.291

TABLE VI - Critical Values of the χ^2 Distribution

Entries in the body of Table VI are the values of χ^2 that cut off the proportion of the sampling distribution that is indicated by the column headings. The row headings indicate the degrees of freedom. For example, when χ^2 has 3 degrees of freedom, a value of χ^2 = 7.815 cuts of the upper .05 of the sampling distribution.

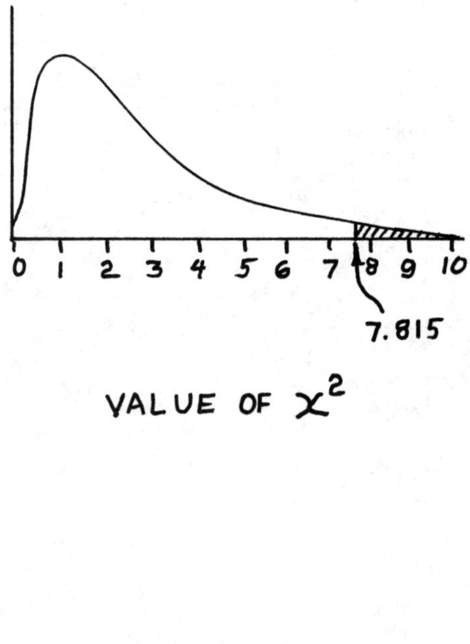

TABLE VI Critical Values in the χ^2 Distribution

	.50	.30	.20	.10	.05	.02	.01	.001
1	.455	1.074	1.642	2.706	3.841	5.412	6.635	10.827
2	1.386	2.408	3.219	4.605	5.991	7.824	9.210	13.815
3	2.366	3.665	4.642	6.251	7.815	9.837	11.345	16.266
4	3.357	4.878	5.989	7.779	9.488	11.668	13.277	18.467
5	4.351	6.064	7.289	9.236	11.070	13.388	15.086	20.515
6	5.348	7.231	8.558	10.645	12.592	15.033	16.812	22.457
7	6.346	8.383	9.803	12.017	14.067	16.622	18.475	24.322
8	7.344	9.524	11.030	13.362	15.507	18.168	20.090	26.125
9	8.343	20.656	12.242	14.684	16.919	19.679	21.666	27.877
10	9.342	11.781	13.442	15.987	18.307	21.161	23.209	29.588
11	10.341	12.899	14.631	17.275	19.675	22.618	24.725	31.264
12	11.340	14.011	15.812	18.549	21.026	24.054	26.217	32.909
13	12.340	15.119	16.985	19.812	22.362	25.472	27.688	34.528
14	23.339	16.222	18.151	21.064	23.685	26.873	29.141	36.123
15	14.339	17.322	19.311	22.307	24.996	28.259	30.578	37.697
16	15.338	18.418	20.465	23.542	26.296	29.633	32.000	29.252
17	16.338	19.511	21.615	24.769	27.587	30.995	33.409	40.790
18	17.338	20.601	22.760	25.989	28.869	32.346	34.805	42.312
19	18.338	21.689	23.900	27.204	30.144	33.687	36.191	43.820
20	19.337	22.775	25.038	28.412	31.410	35.020	37.566	45.315
21	20.337	23.858	26.171	29.615	32.671	36.343	38.932	46.797
22	21.337	24.939	27.301	30.813	33.924	37.659	40.289	48.268
23	22.337	26.018	28.429	32.007	35.172	38.968	41.638	49.728
24	23.337	27.096	29.553	33.196	36.415	40.270	42.980	51.179
25	24.337	28.172	30.675	34.382	37.652	41.566	44.314	52.620

Table VI (cont'd)

	.50	.30	.20	.10	.05	.02	.01	.001
26	25.336	29.246	31.795	35.563	38.885	42.856	45.642	54.052
27	26.336	30.319	32.912	36.741	40.113	44.140	46.963	55.476
28	27.336	31.391	34.027	37.916	41.337	45.419	48.278	56.893
29	28.336	32.461	35.139	39.087	42.557	46.693	49.588	58.302
30	29.336	33.530	36.250	40.256	43.773	47.962	50.892	59.703

APPENDIX II

Computer Exercises

There can be little doubt that the most effective way to acquire a firm grasp of a given statistical concept or procedure is to discover for oneself the way in which its numerical values change as various conditions are changed. In most cases, however, the calculations required to carry out an exercise of this sort are so time consuming and tedious that even the most interested of students is soon discouraged. The computer exercises described in this appendix were written to resolve this problem. More specifically, they were designed to provide the student with an efficient way to explore the numerical properties of the major statistical concepts and procedures that have been described in this book. These exercises are available on a 5-1/4" floppy disk that can be purchased from the Bryn Mawr College Bookshop. Inquiries should be addressed to: The Bookshop, Bryn Mawr College, Bryn Mawr, PA 19010. Two versions of the disk are available. 1) A version that can be run on a TRS 80 Model III or Model IV computer with a disk drive and 48K of memory. 2) A version suitable for the IBM PC with a disk drive or one of its clones. In both cases, a printer with a continuous paper feed is required as neither version can run successfully without it. The instructions which follow are intended for the student who has acccess to a TRS 80 computer but who may not have ever previously run a program.

In order to run a given program you must first turn on the printer and then turn on the computer and wait a few seconds until the disk drives stop running. If your computer and printer are plugged into a switched power strip you may turn on both units simultaneously by turning the power strip on. Next insert your disk into the bottom drive. In doing so, take care that you do not force or bend the disk as it can be easily damaged. Once the disk has been inserted and you have gently closed the door you are ready to set the computer into operation. Press the orange button directly below the door. The computer will immediately read the disk and will ask a variety of questions which you must answer. First the computer will ask you to enter the date. You respond by typing the appropriate date from the keyboard. For example, 02/14/84. Next press the white key entitled ENTER. You may use either the large ENTER key or the small one on the right. The computer will then ask for the time of day. At this point you can either look at your watch and enter the time or just press the ENTER key. Once you have entered the date and time, the computer will whirr a bit and then say TRSDOS Ready. At that point you should write the word BASIC, and press 'ENTER'. Again the computer will whirr some and it will come up with a series of questions. For each of these questions you need only press ENTER (i.e., you need not actually answer the question). Eventually the screen will show an arrow pointing to the left and a flashing white box. At this point, you should type in LOAD and type in the name of the program you wish to run. (Be sure to include the quotation marks around the name.) Then press the ENTER key again. The computer will proceed to load the program. When the program is loaded type RUN and then press the ENTER key. The computer will now run the program. However, before you run any of these programs there are two additional points that need to be clarified.

First. By providing a printout of all of the important information, each program will give you a permanent record of the exercise so that you can study it at your leisure. Since several of the exercises relate to each other in various ways, it will be important to save all of them in a single convenient location.

Second. It is important to note at the outset that ocasionally a number that the computer generates will be so close to zero that its expression requires more than 6 digits beyond the decimal point. When this occurs the computer will present the number in an exponential format and indicate that it is doing so by printing an "E" along with the number. For example, the quantity .000000026 written in an exponential format would appear 2.6 E -08. This should be understood to mean 2.6 times 10^{-8}.

When you have finished running your exercises on a given day the computer should be turned off. To do so you should first gently remove the disk and replace it in its jacket. Then turn off the computer and finally turn off the printer. If both units are plugged into a switched power strip you can turn off both units simultaneously by throwing the switch on the power strip to off.

Computer Exercises on the Transformation of Scores

These notes accompany the computer program entitled "TWELVE". The program is designed to enable you to observe what happens to the mean, variance, and standard deviation of a distribution of scores, when these scores are transformed by adding and/or multiplying each by a given constant. Pages 25-28 of the textbook provide a detailed discussion of these effects. By loading and running "TWELVE" you can explore these effects for yourself.

For your first run you should add a number of your choice to each item in the distribution. Note how this transformation influences the mean of the distribution while leaving the variance (and standard deviation) unchanged.

Next, try multiplying every item in the distribution by a value of your choice. Take note of how this procedure multiplies the mean and the standard deviation by the value and how it multiplies the variance by the square of the value.

Finally, try both adding the first value and multiplying by the second value. Note how the mean, variance and standard deviation of the distribution are effected.

To obtain the new mean you must add the first value to the original mean and then multiply the resultant by the second value. To obtain the standard deviation of the doubly transformed scores you need only multiply the original standard deviation by the second value. Finally, to obtain the variance of the doubly transformed scores, you multiply the variance of the original scores by the square of the second value.

When you have completed the exercise using values of your own choice, you should rerun the program (by typing RUN and hitting the ENTER key). This time, however, you should enter values for addition and for multiplication that will yield a doubly transformed distribution with a mean of zero and a standard deviation of one. In short, you should convert each original score into a Z score.

First, you should examine what happens to the mean, variance and standard deviation of a distribution when you subtract the mean of the distribution from every score. This can be accomplished by adding (minus the mean) to each score. Note that when you do so, the new mean is zero (or a number close to it) and the variance and standard deviation are unchanged.

Next, you will be asked to enter a value by which each score will be multiplied. To divide each original score by the standard deviation of the distribution, you must first use a calculator to compute the reciprocal of the standard deviation (e.g., 1/standard deviation).

If you then multiply every item in a distribution by the reciprocal of the standard deviation, you will produce the same effect as if you had divided every item in the distribution by the standard deviation. Note that when you do so, the new mean will be equal to the old mean divided by the standard deviation. The new variance will equal one (or a number very close to one), and the new standard deviation will also be one (or a number that is close to one)

Finally, you should apply both transformations to every original score (e.g., both subtract the mean and divide by the standard deviation). Note that when you do so, the new mean will be zero and the new standard deviation (and variance) will be one (or numbers that, due to rounding errors, are quite close to one).

Computer Exercises on the Sampling Distribution of the Mean

This materal is designed to accompany the computer program entitled "TEN". This program will enable you to approximate a sampling distribution of the statistic $\bar{X} = \frac{\sum X}{n}$. In doing so it will provide an opportunity to examine, in an empirical fashion, the three factors that together comprise the Central Limit Theorem. As noted on pages 46-60 in the textbook, this theorem describes what happens when one draws an infinite number of random samples, each of size n, from a population with mean = μ_X and variance = σ_X^2. If one calculates the statistic $\bar{X}$ on each sample and then makes a frequency distribution of the $\bar{X}$'s one will find that:

1. $\mu_{\bar{X}} = \mu_X$
2. $\sigma_{\bar{X}}^2 = \frac{\sigma_X^2}{n}$
3. When the population is normal, the sampling distribution of $\bar{X}$ will be normal. When the population is not normal, the sampling distribution of $\bar{X}$ will nonetheless approximate a normal distribution when n (the size of the samples) is large.

To carry out these exercises you should follow the instructions for running a disk program. In this case, you should run the program entitled "TEN". In your initial run you should elect to draw samples of size n = 1.

When the computer has drawn 100 samples and calculated their means, you should observe that the mean of these means is a number that is fairly close to 5.00 and that the S square of the sample means is fairly close to 6.67/1 = 6.67.

When you have completed the initial run, cut off the paper, and enter "RUN" again. This time, however, you should elect to draw samples of size n = 3.

Again, notice that the mean of the sample means is, once more, a number that is reasonably close to 5.00 (which is the mean of the population). This illustrates Part I of the Central Limit Theorem.

You should also take note that the S square of the sample means is a number that is fairly close to the quantity 6.67/3 = 2.2233. This illustrates Part 2 of the Central Limit Theorem.

Again, you should cut off the paper and enter "RUN". This time you should draw samples of size n = 12.

Notice, once more, that the mean of the sample means is a number fairly close to 5.00. Also notice that the S square of the sample means is a number that is fairly close to 6.67/12 = .55583.

Finally, you should compare the several histograms. When you do so notice how the histogram when n = 1 is approximately rectangular. In other words, notice how it approximates the shape of the population from which the samples were drawn. Also notice how the histograms for samples of sizes n = 3 and n = 12 begin to approximate normal distributions. In short, you should note how the histograms illustrate the third aspect of the Central Limit Theorem. By how you should, of course, also be able to see how they also illustrate parts 1 and 2 of the Central Limit Theorem.

Computer Exercises on the Sampling Distribution of Z Scores for Sample Means

These exercises are designed to elaborate the material on pages 57-70 of the textbook. In particular they will enable you to examine what happens when the statistic $Z_{\bar{X}} = \dfrac{\bar{X} - \mu_h}{\sqrt{\dfrac{\sigma_x^2}{n}}}$ is used to determine whether or not to reject the statistical (i.e., null) hypothesis that the population from which a given random sample was drawn has a mean = μ_h = 5.00.

To carry out these exercises you should load and run the program entitled "FOURTEEN". For your initial run, when asked, you should elect to draw samples of size n = 6. Also, when asked, you should elect to draw your samples from a population with a mean of 5.00.

As with earlier programs, you should understand that the expression Z(M) is the computer's way of printing $Z_{\bar{X}}$.

As indicated in the textbook, $Z_{\bar{X}} = \dfrac{\bar{X} - \mu_h}{\sqrt{\dfrac{\sigma_x^2}{n}}}$ is a statistic (e.g., a measure on a sample) that is used to test the hypothesis that the sample was drawn from a population with mean equal to μ_h. When the hypothesis tested is true, the sampling distribution of Z(M) will have a mean of zero and a standard deviation of one. Moreover, if the population is normal, the sampling distribution of Z(M) will be normal. If the population is not normal, the sampling distribution of Z(M) will nonetheless tend to become normal as sample size increases.

When you have completed the initial run on "FOURTEEN", tear off your printout and study it. Notice that the mean of the illustrative sample is a value that is reasonably close to 5 (e.g., the mean of the population) and that the value of Z(M) calculated on the illustrative sample is reasonably close to zero. Use a calculator to determine for yourself exactly how the value of Z(M) was obtained. Notice that to calculate Z(M) you must know the value of = the variance of the population = 6.67.

Next you should examine the values of Z(M) for the 100 samples (each of size n = 6). Notice that most values are near zero and that extreme values (e.g., values larger that +2.5 or less that -2.5) rarely occur. Notice also that the mean of the 100 values is a number that is quite close to zero.

Next you should rerun the program (by typing RUN and hitting ENTER). This time you should enter 8 when asked to indicate the mean of the population from which the samples will be drawn. As in the prior run, however, your sample size should be 6.

When you have completed the run, tear off the printout and compare it to the one obtained earlier. First notice that in the earlier run you were testing the hypothesis that the population mean = 5 when the hypothesis was, in fact, true (e.g., the samples were drawn from a population with a mean of 5). In the second run you were again using the statistic

to test the hypothesis that the population mean was 5, but in fact the samples were being drawn from a population with a mean of 8. In other words, you were calculating

$$Z(M) = \frac{\bar{X} - 5}{\sqrt{\frac{6.67}{6}}}$$

on each of your samples, but the samples were coming from a population in which the mean was 8, rather than 5.

Notice how this affected your results. First, as you might have expected, the mean of your illustrative sample was probably closer to 8 than to 5. Second, the

Z(M) calculated on the illustrative sample was probably closer to 3 than it was to zero. Look at the array of 100 Z(M)s and at the histogram. Notice that when the hypothesis tested is false, and the samples come from a population with a mean of 8 instead of 5, the mean of the values of Z(M) tends to approximate

$$\frac{8 - 5}{\sqrt{\frac{6.67}{6}}} = \frac{3}{1.054} = 2.845$$

Considered together, the two printouts should enable you to better understand how the statistic Z(M) is distributed when the hypothesis tested is true, and how it is distributed when that hypothesis is false by a given amount.

Try rerunning the program using larger or smaller samples, and when the population is = 5.00 (e.g., the hypothesis tested is true) as well as when it is false by larger and smaller amounts. By doing so you should be able to develop an appreciation of how the statistic $Z_{\bar{X}}$ functions in a variety of circumstances.

Computer Exercises on the Sampling Distribution of S Square

This material is designed to accompany the computer program entitled "ELEVEN". This program will enable you to approximate a sampling distribution of the statistic $S_x^2 = \frac{\Sigma(X - \bar{X})^2}{n - 1}$ (called S square). In doing so it will provide an opportunity to examine, in an empirical fashion, what happens when one draws a large number of random samples of size n, and calculates S^2 on each. The textbook describes the sampling distribution of S square in Chapter 6 on pages 93-115.

To carry out these exercises you should follow the instructions for loading and running a disk program. In this case, you should load and run the program entitled "ELEVEN". For your initial run you should elect to draw samples of size n = 2.

When the computer has drawn 100 samples and calculated S square on each, you should observe that the mean of these S squares is a number that is fairly close to 6.67.

When you have completed the initial run, cut off the paper and enter "RUN" again. This time, however, you should elect to draw samples of size n = 12.

Notice that the mean of the sample S squares is, once more, a number that is reasonably close to 6.67 (which is the variance of the population from which the samples were drawn). This illustrates the proposition that the statistic $S_x^2 = \frac{\Sigma(X - \bar{X})^2}{n - 1}$ is an unbiased estimate of σ_x^2.

It is perhaps of interest that (as with the sampling distribution of $\bar{X}$), the variance of the sampling distribution of S^2 is inversely proportional to the size of the samples, but the exact relationship to the size of the samples, but the exact relationship to n is not so simple as for $\overline{X}$. Fortunately, however, for present purposes, we need only concern ourselves with the mean of the sampling distribution of S square and with gaining insight into how the shape of the distribution of S square changes as sample size increases.

In this regard, you should notice how, when n = 2, the histogram of the 100 values of S_x^2 is quite skewed, but that when n = 12, the histogram begins to approximate a normal distribution.

Computer Exercises on the Sampling Distribution of t

These exercises are designed to elaborate the material in Chapter 7 (pages 116-128) of the textbook. In particular they will enable you to examine what happens when the statistic

$$t_{df=n-1} = \frac{\bar{X} - \mu_h}{\sqrt{\frac{s_x^2}{n}}}$$

is used to determine whether or not to reject the statistical (i.e., null) hypothesis that the population from which a given random sample was drawn has a mean = μ_h = 5.00.

To carry out these exercises you should load and run the program entitled "THIRTEEN". For your initial run, when asked, you should elect to draw samples of size n = 6. Also, when asked, you should elect to draw your samples from a population with a mean of 5.00.

As indicated in the textbook, $t_{df=n-1} = \frac{\bar{X} - \mu_h}{\sqrt{\frac{s_x^2}{n}}}$

is a statistic (i.e., a measure on a sample) that is used to test the hypothesis that the sample was drawn from a population with mean equal to μ_h . When the hypothesis tested is true, the sampling distribution of t will have a mean of zero. However, even if the population is normal, the sampling distribution of t will not be normal, though it will tend to become normal as sample size increases.

When you have completed the initial run on "THIRTEEN", tear off your printout and study it. Notice that the mean of the illustrative sample is a value that is reasonably close to 5 (e.g., the mean of the population) and that the value of t calculated on the illustrative sample is reasonably close to zero. Use your calculator to determine for yourself exactly how the value of t was obtained. Notice that to calculate t, you must use the S Square calculated on the sample where

$$s_x^2 = \frac{\Sigma(x - x)^2}{n - 1}$$

Now examine the values of t for the 100 samples (each of size n = 6). Notice that most values are near zero and that extreme values (e.g., values larger that +2.5 or less that -2.5) rarely occur. Notice that the mean of the 100 values is a number that is quite close to zero. Finally, you should compare the histogram obtained here to the histogram that was generated when samples of size n = 6 were drawn from a population where the population mean was 5.00 and

$$z_{\bar{x}} = \frac{\bar{X} - \mu_h}{\sqrt{\frac{\sigma_x^2}{n}}}$$

was calculated on each sample. Notice that when compared to the sampling distribution for t, the sampling distribution for $z_{\bar{x}}$ has fewer extreme values.

Next you should rerun the program (by typing RUN and hitting ENTER). This time you should enter 8 when asked to indicate the mean of the population from which the samples wil be drawn. As in the prior run, however, your sample size should be 6.

When you have completed the run, tear off the printout and compare it to the one obtained earlier. First notice that in the earlier run you were testing the hypothesis that the population mean = 5 when the hypothesis was, in fact, true (e.g., the samples were drawn from a population with a mean of 5). In the second run you were again using the statistic t to test the hypothesis that the population mean was 5, but in fact the samples were being drawn from a population with a mean of 8. In other words, you were calculating

$$t_{df=5} = \frac{X - 5}{\sqrt{\frac{s_x^2}{6}}}$$

on each of your samples, but the samples were coming from a population in which the mean was 8, rather than 5.

Notice how this affected your results. First, as you might have expected, the mean of your illustrative sample was probably closer to 8 than to 5. Second, the t calculated on the illustrative sample was probably closer to 3 than it was to zero. Look at the array of 100 Ts. Notice that when the hypothesis tested is false, and the samples come from a population with a mean of 8 instead of 5, the mean of the values of t tends to approximate

$$\frac{8 - 5}{\sqrt{\frac{6.67}{6}}} = 2.845$$

Considered together, the two printouts should enable you to better understand how the statistic t is distributed when the hypothesis tested is true, and how it is distributed when that hypothesis is false by a given amount.

Try rerunning the program using larger or smaller samples, and when the population is = 5.00 (e.g., the hypothesis tested is true) as well as when it is false by larger and smaller amounts. By doing so you should be able to develop an appreciation of how the statistic t functions in a variety of circumstances.

Computer Exercises on Correlation and Regression

These notes accompany the program entitled "NINE". The program is designed to permit you to observe many of the factors that, considered together, enable one to obtain a quantitative measure of the degree to which paired measures on each of a set of N items are co-varying. Chapter 11 (pages 162-208) of the textbook provides a detailed account of the concepts that underly correlation and regression. The program entitled "NINE" will enable you to examine many of those concepts in a direct empirical fashion.

When you have loaded and run "NINE", the computer will ask you how many items (or subjects) you wish to test. While you can, if you choose, enter any number from 2 up, for the first run you should enter 8. The computer will then proceed to present a sequence of 8 numbers, each of which is to be conceptualized as a measure on the same variable (for example, the weight) of each of a set of N = 8 subjects. For each subject you should enter a number (of your own choice) to represent a measure on another variable (for example, height) for each of these subjects. In doing so, recognize that the paired numbers can be measures on any two variables and that how an item is defined is determined by practical considerations. Thus, for example, the items might be defined as years in a record book. The X measure might be the amount of rainfall per year, whereas the Y measure might be the lowest temperature recorded in those same years. Alternatively, the items might be conceptualized as students, in which case, the X measure might represent the hours studied prior to a given exam and the Y measure might represent the numerical score achieved on that exam. In such a case, on might wish to determine the degree to which the scores on the exam co-vary with the hours studied.

For the first few runs on this computer program, imagine that each X measure presented to you represents the number of hours a given student studied prior to a given exam. For each student you are to enter a two digit number (from 0 to 9) to represent the number of correct answers for the student on the exam.

For the first run, if the X value is high (7, 8, or 9), you should enter a large number. If the X value is low (0, 1, 2, or 3), enter a small number. When your entries have been completed, the computer will list them and perform a variety of calculations to illustrate various aspects of the concepts of correlation and regression.

First, the computer will calculate the mean and variance of the X measures and of the Y measures. Next, it will convert each of the X into Z scores and it will do the same for the Y measures. Use your calculator to check for yourself that a given

$Z_X = \frac{X - \mu_X}{\sigma_X}$ and that a given $Z_Y = \frac{Y - \mu_Y}{\sigma_Y}$

Also check that the mean of the Z_X s is a number that (given rounding errors) approximates zero and that the variance of the Z_X s is a number that (given rounding errors) approximates one. You should also use your calculator to do the same for the Z_Y scores. Again, the mean of the Z_Y scores should be near zero and the variance should be near one.

The printout also shows the product of the Z_X and Z_Y scores for each item. Use a calculator to check one or two of these products to insure that you understand how they were obtained. The coefficient of correlation is equal to the mean of the Z score products. Use your calculator to check this assertion for yourself.

The printout will next show the parameters b and a in the regression equation used to predict Y from X. See pages 180-183 in the textbook for an introduction to regression equations.

The printout will next show again the values of X, but now, using the regression equation (with its previously calculated parameters (b and a), it will calculate a predicted value of Y for each value of X. It will also show the observed values of Y (e.g., the values you entered), and it will calculate the difference between the observed and predicted values of Y.

The computer will next calculate and the printout will show the variance of the predicted values of Y ($\sigma^2_{\tilde{Y}}$ in the textbook) and the variance of the differences (σ^2_{err} in the textbook). Use your calculator to calculate these variances for yourself.

As noted on page 196 of the textbook, for any set of bivariate data $\sigma^2_Y = \sigma^2_{\tilde{Y}} + \sigma^2_{err}$. Use your calculator to check this assertion for yourself.

You should also use your calculator to check the assertion that $r^2 = \frac{\sigma^2_{\tilde{Y}}}{\sigma^2_Y}$ and $r^2 = 1 - \frac{\sigma^2_{err}}{\sigma^2_Y}$

That is, you should square the mean of the Z score products and compare this value to the value you get when you divide the variance of the predicted Ys by the variance of the Ys and when you subtract $\frac{\sigma^2_{err}}{\sigma^2_Y}$ from one.

Finally, before leaving this exercise, you should use a calculator to determine the value of the sum of the squared Z scores. When you do so, you should obtain a value that (within rounding error) is very close to N (the number of scores). (A proof showing why this happens is given on page 171 of this book.)

You should also use a calculator to determine the mean of the predicted Ys and the mean of the differences. You should find that the former is (within rounding error) equal to the mean of the Y values, while the latter approximates zero.

When you have completed all of the above, you should rerun "NINE", but this time you should enter small numbers when X is 7, 8, or 9 and you should enter large numbers when X is small, (e.g., 0, 1, 2 or 3). As you will discover by redoing the entire exercise; data of this sort yield negative coefficients of correlation.

Finally, if you again rerun "NINE" but this time enter your Y values without looking at the X values, you will be able to observe what happens to the coefficient of correlation and to regression when there is a near zero relationship between the X and the Y values for each item.

Before leaving the topic of correlation and regression, it should be mentioned that there is much to be gained by plotting the various bivariat distributions that are generated in the course of these exercises. Try plotting the scatterplot for the original X and Y values and then try plotting the same data after each item has been converted to a Z score. When plotted in Z score form, the tangent of the slope of the regression line should prove to equal r.

Computer Exercises on One-Way Analysis of Variance

This material is intended to accompany the computer program entitled "EIGHT". Use of the program assumes knowledge of the sampling distribution of means and the central limit theorem (covered in computer program "TEN" and on pages 46-57 in the textbook). It also assumes knowledge of the material covered in Chapter 6 of the textbook (pages 93-115) and the concepts covered in the computer program labeled "ELEVEN". Finally, use of the program assumes knowledge of the basic concept of treatment effects, covered on pages 251 to 253 of the textbook.

The program is designed to clarify the logic of a one-way analysis of variance by conducting a one-way analysis on a set of randomly selected items and indicating the values obtained at each stage of the computations. The logic of the one-way analysis of variance is treated in detail in Chapter 14 (pages 232-253) of the textbook. By working through the program entitled "EIGHT" you should be able to clarify any ambiguities that remain after you have read Chapter 14.

An Example of the One Way Analysis of Variance When The Null Hypothesis is True (i.e., The Treatment Effect for All Conditions is Zero)

When you run "EIGHT" the computer will ask you to enter the number of conditions. For the initial run you should type in 4 and then hit the 'ENTER' key. Next the computer will ask you to enter the number of observations per condition. You should enter 8 for this initial run. The computer will then proceed to draw and list 4 sets of 8 randomly selected items. It will then ask you if you wish to add a treatment effect to the items in each condition. For this run you should either press enter for each query or enter a 0.

The printer will now list the entries, after adding the treatment effects you designated. In this instance, since the treatment effects were always zero, the post-treatment entries will be identical to the initial entries. As indicated in the printout, the initial set of entries represent randomly selected items from a population with a mean = 5.00 and variance = 6.67. Since, in the present run no treatment effects are added, the post-treatment items represent the state of affairs that prevails when one performs an analysis of variance, and the null hypothesis is true. In other words, the initial run simulates the conditions that prevail when one performs an analysis of variance on data where the observed differences among the means of the 4 conditions are in fact solely due to random variations in sampling.

You, of course, would not be privileged with this insight in an actual study. Accordingly, you would have to carry out the analysis to decide whether or not you could reject the null hypothesis. By introducing zero treatment effects in the present exercise, you have the opportunity to observe what happens when an Analysis of Variance is conducted on data where the null hypothesis is true.

By then repeating the analysis using the same randomly selected items as before, but now introducing various treatment effects, you will be able to observe how the analysis changes when the null hypothesis is false, and hence should be rejected. For the time being, however, let's consider the four means listed in the initial printout. If this was a real experiment or a survey instead of a simulation, these means would be your primary interest and you would probably want to plot them in a bar graph. Whether you did so or not, the question that would need to be resolved, and the reason for the statistical analysis, would be whether or not the observed differences among the condition means was greater than would be expected on the basis of chance.

The statistical measure of these observed differences is the S^2 for condition means. We know from the Central Limit Theorem that if we take samples of size n = 8 from a population with a mean of 5 and a variance of 6.67 (as we have in the present instance), the mean of the sampling distribution of means will equal 5.00 and the variance of the sampling distribution of means will equal 6.67/8 = .83375.

In this instance we have (in effect) a sample of 4 items (i.e., 4 means) from such a sampling distribution of means. Accordingly, the mean of this sample of 4 means (the grand mean) should be a number that is fairly close to 5.00. Similarly, the S Square of the four condition means should be a number that is fairly close to 6.67/8 = .83375. Check the printout to see that both of these assertions are true.

In the analysis of variance, as noted in Chapter 14 of the textbook, we use the observed variation within the conditions to help decide whether or not the observed variations between the means of the conditions should be attributed to chance. We know, from our studies of the sampling distributioon of the statistic S Square, that the S Squares of each of the four grops is an unbiased estimate of the variance of the population from which the samples were drawn. Notice in the printout that, give and take some expected variation (recall your prior studies of the sampling distribution of S^2_X), the S Squares within the conditions are all reasonably close to 6.67, as is the average of these S Squares.

In the Analysis of Variance Table (at the bottom of the printout) the Mean Square Within is equal to the average of the within condition S Squares. The Mean Square Between, on the other hand, is equal to the S Square of the condition means, multiplied by the size of the samples (in this case 8) upon which the condition means are based. The logic of this proposition bears repeating:

> Since we have added no treatment effects we know that the S Square of the condition means is an unbiased estimate of $\sigma^2_{\bar{X}}$ = 6.67/8 = .83375. Accordingly, if we multiply the observed S Square of the condition means by n = 8, we convert it to an unbiased estimate of the variance of the population from which the samples were drawn.

Notice how the Mean Square Between, like the Mean Square Within, is fairly close to 6.67. Of course it will seldom be exactly 6.67, and on occasion, it will depart appreciably from 6.67. But these are the random variations we expect, and as discussed in the text, if you always test at the $\alpha = .05$ level of significance, and if your treatments never have an effect, once out of twenty times (on the average) you will obtain a result that will lead you to reject the null hypothesis.

Notice also, that the value of F, in the Analysis of Variance Table at the bottom of the page is equal to the Mean Square Between divided by the Mean Square Within. You should also notice that the between Sum of Square is equal to the Mean Square Between times its degrees of freedom. Similarly the Sum of Squares Within is equal to the Mean Square Within times its degrees of freedom. In each of these cases you should use your calculator to check the assertion for yourself.

An Example of the One Way Analysis of Variance When The Null Hypothesis is Not True (i.e., When There are Non Zero Treatment Effects

For these exercises you should continue with the set of observations that were entered to study the analysis of variance when the null hypothesis was true in all instances. You will have to begin all over again with a new set of data. If you still have the computer turned on and have just completed the analysis of a given set of data under the null hypothesis, you will note that the computer has stopped with a question "Do you wish to examine the analysis of variance on these data when various treatment effects have been added to the items in each cell?" You can answer this question by entering N for no or Y for yes. If you enter N for no, the program will end and you will have to start all over again if you are to examine the same data with and without treatment effects. If, on the other hand, you enter Y the computer will reprint your original set of entries and then ask you what effects you wish to add. At this point, you can enter whatever effects you wish to study. A good way to go about this is to try to arrange to keep the effects that you add so that they sum to zero across the several treatment conditions. By doing so, you will leave the grand mean as it was and in this way make it easier to see exactly how the treatment effects influence the outcome of the analysis. Try, for example, adding -4 to each item in column 1, add 1 to the items in column 2, add 2 to the items in column 3, and add 1 to the items in column 4. When the run is completed, compare the printout to the printout obtained for the same set of observations when there were no treatment effects. Notice that by introducing treatment effects you cause the MS for conditions to increase, but you do not effect the MS Within. Accordingly, F becomes large - a condition that would usually lead you to reject the null hypothesis.

As with previous exercises, it will be instructive to try various sample sizes, numbers of conditions, and various treatment effects. By studying the several printouts and comparing them to each other, you should be able to develop a good appreciation of how the One-Way ANOVA enables you to make inferences about statistical significance in a wide variety of circumstances.

Computer Exercises on Two-Way Analysis of Variance

Use of the program assumes knowledge of the Central Limit Theorem (covered on pages 46-60 of the textbook). It also assumes knowledge of One-Way Analysis of Variance (covered on page 232-253 and of Two-Way Analysis of Variance covered on pages 254-279). In addition, one should be familiar with the concept of treatment effects (on pages 251-253 in the same text). The program is designed to clarify the logic of factorial designs by carrying out a two-way analysis of variance on a set of randomly selected items and indicating the values that are obtained at each stage of the computations. The name of this program is "SEVEN". When you run it the computer will ask you a series of questions. The initial question will be how many row conditions you wish to test. In response to this question you may enter any number from 2 up. The computer will then ask how many column conditions you wish to test. You may now enter either 2, 3, or 4. The limination here is caused by the need to keep the data in an easily read format. Finally the computer will ask how many observations you wish to have in each cell. Again you can enter any number from 2 up. Once you have set up the parameters for the analysis, the computer will proceed to draw items at random from a population with a mean of 5.0 and a variance of 6.67 and enter them into an array containing the rows and columns and numbers of observations per cell that you have elected to test.

Once the array has been printed, the computer will ask you what row, column, and interaction effects you may wish to add. This time You may enter any numbers you wish.

The computer will next provide a printout of the row effect that you elected to add to the items in Row 1, Row 2, etc. It will also indicate the column effects you elected to add to the items in Columns 1, 2, etc., and it will indicate the interaction effects you elected to add to Cells 1 to C in each row. The computer will then add these effects to the items in each of the cells and print out the new values.

The program will next direct the computer to carry out an analysis of variance. However, unlike most analysis of variance programs that have been written for computers, this particular program employs conceptual (rather than computational) formulas in the course of its calculations and what is especially important for instructional purposes, it prints out these calculations as it moves along. Another unique feature of this program is that once it has carried out an analysis on a given set of randomly selected items with a given set of user selected treatments, the computer can, if instructed to do so, repeat the analysis on the same original items but with different user specified treatment effects. This enables the user to learn how the the two-way analysis of variance functions in various hypothetical situations.

In general, the program can deal with almost any set of conditions you might wish to study, but you should understand that this is only a small computer and that it is possible to set it a problem that exceeds its memory capabilities. If this has happened the computer will proceed with the calculations until it runs out of memory, whereupon it will tell you so. At this point, you will have no choice but to press the orange (reset) key (this key is directly under the disk drives) and begin again.

An Example of a Two-Way Analysis of Variance When There Are no Row, Column, or Interaction Effects (e.g., When the Null Hypothesis is True in all Instances)

For this exercise you should load and run "SEVEN" and when the computer asks you "How many rows?", enter 2. When it asks "How many columns?", enter 3, and for observations per cell enter 4. The computer will then take its random samples and when it prompts you as to what row, column and treatment effects you wish to add, enter 0's or simply push the ENTER key in response to each of its questions. When the computer finishes its run, tear off the printout and study it in conjunction with the comments which are listed below.

The first thing to notice in the printout is that the mean for each of the cells is reasonably close to the mean of the population. This should make sense to you. After all, each of the items in the array was drawn from a population with mean = 5.0. Thus, each of the cell means is based on a random sample of 4 items from a population with a mean of 5.0. The same will be true of the row means, the column means, the corrected cell means, and the grand mean. Each is an unbiased estimate of the mean of the population (e.g., 5). One would expect, however, that means based on large numbers of items (the grand mean is based on RCn = 24 items) should be closer to 5 than means based on fewer items (a given cell mean is based on only n = 4 items).

Let's now turn our attention to the several kinds of variability that the various means exhibit. The Central Limit Theorem tells us that the variance of a given sampling distribution of means is equal to the variance of the population divided by the size of the samples upon which the means are based. These considerations imply that if the null hypothesis is true (i.e., there are no treatment or interaction effects) the S square for the original cell means is an unbiased estimate of the variance of the sampling distribution of means for samples of size 4. This implies that if we multiply the obtained S square for the original cell means by the number of items (e.g., 4) on which each mean is based, we convert it into an unbiased estimate of the variance of the population from which the samples were drawn. Try carrying out this calculation to see how close the estimate is to 6.67.

Regardless of how close the estimate happens to be, it is important to recognize that this estimate appears in the Analysis of Variance Table as the Sum of Squares for Sub Total divided by degrees of freedom for Sub Total. Try it and examine the analysis of variance table (at the bottom of the printout) to see if it doesn't work out.

Let's next examine the S square for the individual cells. Use your calculator to check that these are in fact the S squares of the items that appear in each of the several cells. Each of these S squares is an unbiased estimate of the variance of the population. Accordingly, each of them should be relatively close to 6.67 and the average of them (indicated just above the analysis of variance table on the bottom of the page) should also be reasonably close to 6.67. Notice that the average of the within cells S squares is equal to the mean square within.

You should also notice that the sum of squares for within is the mean square within times the number of degrees of freedom that it contains. In this instance we have 6 groups with 4 items in each. In calculating each of the S squares, however, we lost a degree of freedom because we had to use the mean of the cell to estimate the mean of the population. Accordingly, there are only 3 degrees of freedom in each of the 6 S squares that contribute to the S square within.

Now, let's examine the means for the rows. The printout shows the mean for Row 1, the mean for Row 2 and it reminds you that each row mean is based on the data from C x n observations. In this run C x n = 12. The printout also shows the S square for the row means. From the Central Limit Theorem you know that this number is an unbiased estimate of the variance of the sampling distribution of means for samples of size 12 when the samples are drawn at random from a population with a variance of 6.67. If you multiply the S square for the row means by the size of the samples on which those means are based, you convert it from an estimate of the variance of a sampling distribution of means to an estimate of the variance of the original population from which the samples were drawn. Multiply the S square for the row means by the quantity C x n = 12 and you will obtain a number that will equal the mean square for rows. If you multiply the mean square for Rows by its degrees of freedom (e.g., 1), you will obtain the Sum of Squares for Rows.

Next, look at the means for columns 1-3. The printout reminds you that each of the column means is based on R x n (which in this case) equals 8 oservations and it shows the obtained value of S square for the column means. When the null hypothesis is true (as it is in this instance), the S square for column mean is an unbiased estimate of the variance of the sampling distribution of means when the samples are drawn from a population with a variance of 6.67 and when the samples contain R x n = 8 observations. Again, if we multiply the S square for column means by the number of items on which each mean is based, we obtain an estimate of the population variance. This quantity is called the mean square for columns. Use your calculator to check this for yourself. If you now multiply the mean square for columns by the degrees of freedom that it contains, you will arrive at the quantity called the sum of squares for columns. Again check this assertion with your calculator.

The corrected cell means are of special interest. To understand how they are derived it is necessary to recognize that the observed differences among the row means and among the column means can be conceptualized as representing observed effects of the row treatments, of the column treatments and of the interactions among these treatments. In the data we are working with here, we have elected to add no row, column, or interaction effects, but we nonetheless observe that there is variability among the row means and there is variability among the column means. These observed variabilities are conceived to represent observed or empirical effects that, in the present circumstance, are due to random variations. If one subtracts the grand mean from a given row mean, one has a quantitative measure of a given observed row effect. If one subtracts the grand mean from the observed mean of a given column, one has a quantitative measure of a given observed column effect.

Each corrected cell mean is obtained by first calculating the mean for a given cell and then subtracting the observed row effect for that cell (e.g., the row mean for that row minus the grand mean) and also subtracting the observed column effect for that cell (e.g., the column mean for that column minus the grand mean). At this point you should actually carry out these calculations and determine that this is in fact how the corrected cell means were obtained.

An important characteristic of the corrected cell means is that the mean of each row is now equal to the grand mean and that the mean of each column is also equal to the grand mean. Use your calculator to check these statements for yourself.

At this point go back and check the S square for the original cell means. Use your calculator. Notice that the value you obtain if you use the sum of squared deviations of the cell means from the grand mean divided by the number of cell means minus 1 (R x C - 1), your calculations will work out fine. When you calculate the S square of the corrected cell means, however, you encounter a different situation. In this case, if you divide by the number of corrected cell means minus 1, the value you obtain will not agree with the number that the computer has calculated. This is because the corrected cell means have less degrees of freedom than were in the original cell means and in calculating the S square for the corrected cell means the computer took this into account. For the present data, there are only 2 entries in the set of corrected cell means that had to be calculated directly (e.g., by subtracting an empirical row and column effect from a given original cell mean). Once you have calculated the value of any 2 corrected cell means, (they should not both be in the same column), all the rest of the corrected cell means can be obtained by subtraction. This is a consequence of the fact that for corrected cell means, the mean of all rows and columns equal the grand mean. Try it for yourself. Pick two original cell means (each from a different column) calculate their corrected values by subtracting the empirical row and column effects. Then determine the rest of the corrected cell means by appropriately subtracting the values for the two corrected cell means from the grand mean.

Because of the above restrictions an array which has R rows and C columns of corrected means will have only (R - 1) x (C - 1) degrees of freedom - in this instance 2. Try again to calculate the S square for corrected cell means. This time be sure to divide the sum of squared deviations by the appropriate degrees of freedom, namely, 2.

The S square for corrected cell means is an unbiased estimate of the variance of the sampling distribution of means for samples of size n. Accordingly, if you multiply the S square for corrected cell means by 4 (the number of items on which each mean is based), you will obtain another estimate of the variance of the population. This quantity is called the mean square for interaction.

The average of the within cell S squares is reported next. Use your calculator to check it. Take the average of the S squares for the 6 cells. It should equal the quantity labeled as the average of the within cell S squares. This quantity is the mean square within. If you multiply the mean square for within by its degrees of freedom. You will obtain the within sum of squares.

Next, you should direct your attention to the analysis of variance table at the bottom of the page. Once you have the within sum of squares, as well as sums of squares for the rows, columns and interaction, you have all of the information needed to fill out the entire analysis of variance table, that is, of course, assuming you know the degrees of freedom associated with each of these sums of squares. You should check this out for yourself. The values of F shown here reflect the mean square for rows divided by the within mean square, the mean square for columns divided by the within mean square and the mean square for interaction divided by the within mean square. It will be instructive to look up the critical values of F at, for example, $\alpha = .05$ for each of these F's. Since the F's shown here were calculated on data that represent conditions when the null hypothesis is true for each of the tests, it is unlikely that any of the F's you obtain will exceed the critical values listed in the tables. Of course, once out of 20 times, on the average, one will get get a value that exceeds the critical value using the procedures we used here.

The next step in these exercises is to repeat the same kind of analysis that you just completed, one in which there are no row, column, or interaction effects but in which you try different numbers of rows, different numbers of columns, and most importantly, different sample sizes. Notice how as the sample size increases the estimates of the population means and variances become more and more accurate and the F's tend to approximate numbers that are closer to 1, that is, the values of obtained F are less variable than when smaller samples or smaller numbers of rows and columns are used. In all cases, however, the probabilities of getting a significant value of F (a value that exceeds the critical value for the appropriate numbers of degrees of freedom) is always the value that you set for α.

Some Examples of the Two-Way Analysis of Variance When theNull Hypothesis is Not True (e.g., When There are Either Row, Column or Interaction Effects)

For these exercises you can either continue with a set of observations that were entered to study the analysis of variance when the null hypothesis was true in all instances, or you can begin with a new set of data. If you still have the computer turned on and have just completed the analysis of a given set of data under the null hypothesis you will note that the computer has stopped with the question "Do you wish to examine the analysis of variance on these data when various treatment effects been added to the items in each cell. You can answer this question by entering N for no or Y for yes. If you enter N for no, this is the end of the program and you will have to collect a new set of data to run. If, on the other hand, you enter Y the computer will reprint the original set of numbers and then ask you what effects you wish to add. At this point, you can enter whatever effects you wish to study. A good way to go about this is to try to arrange to keep the effects that you add so that they sum to zero for rows, for columns, and for interaction. If you have, for example, 3 columns and wish to add effects that will sum to zero across the columns, you could add, for instance, a -2 to the items in the first column, a 0 to the items in the second column and a +2 to the items in the third column. You will have added effects that will increase the variability of the means of the columns but they will not effect the grand mean nor will they effect the S square within the columns. Moreover, the S square for rows as well as the S square for interaction will be unchanged. Once you have entered a set of column effects, and have seen how things change, try adding row effects. For example, add +2 to the items in the first row and add -2 to the items in the second row. Try adding both row effects and column effects. Finally, try adding interaction effects. Interaction effects can be added without adding row and column effects if you distribute the interaction effects among the several cells so that the interaction effects in each row sum to 0 and the interaction effects in each column sum to 0. For example, if you add -2, 0 and +2 respectively to the items in each of the three cells in row one and then add +2, 0 and -2 to the items in each of the three cells in row two, you will have added interaction effects without having added either row or column effects. The analysis will reflect this fact by exhibiting an enlarged Interaction Mean Square, but the other aspects of the analysis will be unchanged.

Computer Exercises on Two Way Analysis of Variance Using User Generated Data

It should be noted that there is another program to use if you wish to examine the effects of various treatments on data of your own, rather than the computer's choice. This program is entitled "SIX". When you run "SIX" the computer asks you to enter any numbers you choose. Once the numbers have been entered, the computer will give you the same prompts and carry out the same calculations as it does when it generates its own data. Moreover, like the prior program, it will allow you to go back and repeat the analysis on the original data with any new treatment effects you may care to study.

Acknowledgement

The impetus to write this book arose almost 10 years ago when my son Russell was attending Pennsylvania State University and decided to take an introductory course in statistics. I remembered how disappointed I had been when I took my first course in statistics and discovered that in many cases the sources of the formulae and procedures I was studying were never elaborated fully. I found this to be quite dissatisfying and subsequently spent much time and energy in trying to discover these sources for myself.

When I learned that my son would be studying statistics I wanted to provide him with an account of the insights I had gleaned over the years. To do so, I wrote fairly detailed notes on the lectures I was currently presenting at Bryn Mawr College and sent them to him as the semester proceeded. This book is an expansion of those notes.

I owe a special debt to Doris McCullough. She typed the manuscript and worked with me on all phases of its development. Without her encouragement and cheerful willingness to proceed through its numerous revisions, this book would have been abandoned years ago.

I also wish to thank Gray Snowden for writing the portion of the computer programs devoted to the presentation of the histograms. I had puzzled over how to accomplish this but my own skills as a computer programmer had proved too limited for the task. I did, however, write all of the rest of the computer programs and hence am responsible for their weaknesses as well as their strengths.

Finally, I wish to thank my wife, Alice. Her daily work as a historian and labor educator has provided me with a view of effective teaching that I have rarely encountered in my own academic and scientific circles. It has served as a model for my teaching and has set a standard to be approximated as I wrote this book.

The statistical tables in Appendix I are derived from a number of sources.

Table I is adapted from Appendix 2 of R. Clarke, A. Coladarci, and J. Caffrey, Statistical Reasoning and Procedures. Charles E. Merrill Publishing Company, Columbus, Ohio: 1965. It is reprinted by permission of R. B. Clarke.

Table II is adapted from Table A-6 of Wilfrid J. Dixon, Frank J. Massey, Jr., Introduction to Statistical Analysis, 3rd ed. McGraw Hill Book Company, New York: 1969. It is reprinted by permission of McGraw Hill Book Company.

Table III is adapted from Table 31 of E. Pearson & H. O. Hartley, Biometrika Tables for Statisticians, 2nd ed. Cambridge University Press, New York: 1958. It is reprinted by permission of Biometrika Trustees.

Table IV is adapted from Table 31 of E. Pearson & H. O. Hartley, Biometrika Tables for Statisticians, 2nd ed. Cambridge University Press, New York: 1958. It is reprinted by permission of Biometrika Trustees.

Table V is taken from Table III of R. A. Fisher & F. Yates, Statistical Tables for Biological, Agricultural and Medical Research, published by Longman Group Ltd., London: 1974 (previously published by Oliver and Boyd Ltd., Edinburgh) and by permission of the authors and publishers. I am grateful to the Literary Executor of the late Sir Ronald A. Fisher, F.R.S., to Dr. Frank Yates F.R.S. and to Longman Group Ltd., London for permission to reprint a portion of Table III in their book Statistical Tables for Biological, Agricultural and Medical Research (6th Edition, 1974).

Table VI was adapted from Table II of R. A. Fisher, Statistical Methods for Research Workers, Hafner Press, New York: 1973. It is reprinted by permission of MacMillan Publishing Co., Inc.